FREE Test Taking Tips DVD Offer

To help us better serve you, we have developed a Test Taking Tips DVD that we would like to give you for FREE. **This DVD covers world-class test taking tips that you can use to be even more successful when you are taking your test.**

All that we ask is that you email us your feedback about your study guide. Please let us know what you thought about it – whether that is good, bad or indifferent.

To get your **FREE Test Taking Tips DVD**, email freedvd@studyguideteam.com with "FREE DVD" in the subject line and the following information in the body of the email:

a. The title of your study guide.

b. Your product rating on a scale of 1-5, with 5 being the highest rating.

c. Your feedback about the study guide. What did you think of it?

d. Your full name and shipping address to send your free DVD.

If you have any questions or concerns, please don't hesitate to contact us at freedvd@studyguideteam.com.

Thanks again!

TSI Math Study Guide 2019

TSI Math Workbook & 2 Practice Tests for the Texas Success Initiative Assessment

Test Prep Books 2018 & 2019 Team

Table of Contents

Quick Overview

As you draw closer to taking your exam, effective preparation becomes more and more important. Thankfully, you have this study guide to help you get ready. Use this guide to help keep your studying on track and refer to it often.

This study guide contains several key sections that will help you be successful on your exam. The guide contains tips for what you should do the night before and the day of the test. Also included are test-taking tips. Knowing the right information is not always enough. Many well-prepared test takers struggle with exams. These tips will help equip you to accurately read, assess, and answer test questions.

A large part of the guide is devoted to showing you what content to expect on the exam and to helping you better understand that content. Near the end of this guide is a practice test so that you can see how well you have grasped the content. Then, answers explanations are provided so that you can understand why you missed certain questions.

Don't try to cram the night before you take your exam. This is not a wise strategy for a few reasons. First, your retention of the information will be low. Your time would be better used by reviewing information you already know rather than trying to learn a lot of new information. Second, you will likely become stressed as you try to gain large amount of knowledge in a short amount of time. Third, you will be depriving yourself of sleep. So be sure to go to bed at a reasonable time the night before. Being well-rested helps you focus and remain calm.

Be sure to eat a substantial breakfast the morning of the exam. If you are taking the exam in the afternoon, be sure to have a good lunch as well. Being hungry is distracting and can make it difficult to focus. You have hopefully spent lots of time preparing for the exam. Don't let an empty stomach get in the way of success!

When travelling to the testing center, leave earlier than needed. That way, you have a buffer in case you experience any delays. This will help you remain calm and will keep you from missing your appointment time at the testing center.

Be sure to pace yourself during the exam. Don't try to rush through the exam. There is no need to risk performing poorly on the exam just so you can leave the testing center early. Allow yourself to use all of the allotted time if needed.

Remain positive while taking the exam even if you feel like you are performing poorly. Thinking about the content you should have mastered will not help you perform better on the exam.

Once the exam is complete, take some time to relax. Even if you feel that you need to take the exam again, you will be well served by some down time before you begin studying again. It's often easier to convince yourself to study if you know that it will come with a reward!

Test-Taking Strategies

1. Predicting the Answer

When you feel confident in your preparation for a multiple-choice test, try predicting the answer before reading the answer choices. This is especially useful on questions that test objective factual knowledge or that ask you to fill in a blank. By predicting the answer before reading the available choices, you eliminate the possibility that you will be distracted or led astray by an incorrect answer choice. You will feel more confident in your selection if you read the question, predict the answer, and then find your prediction among the answer choices. After using this strategy, be sure to still read all of the answer choices carefully and completely. If you feel unprepared, you should not attempt to predict the answers. This would be a waste of time and an opportunity for your mind to wander in the wrong direction.

2. Reading the Whole Question

Too often, test takers scan a multiple-choice question, recognize a few familiar words, and immediately jump to the answer choices. Test authors are aware of this common impatience, and they will sometimes prey upon it. For instance, a test author might subtly turn the question into a negative, or he or she might redirect the focus of the question right at the end. The only way to avoid falling into these traps is to read the entirety of the question carefully before reading the answer choices.

3. Looking for Wrong Answers

Long and complicated multiple-choice questions can be intimidating. One way to simplify a difficult multiple-choice question is to eliminate all of the answer choices that are clearly wrong. In most sets of answers, there will be at least one selection that can be dismissed right away. If the test is administered on paper, the test taker could draw a line through it to indicate that it may be ignored; otherwise, the test taker will have to perform this operation mentally or on scratch paper. In either case, once the obviously incorrect answers have been eliminated, the remaining choices may be considered. Sometimes identifying the clearly wrong answers will give the test taker some information about the correct answer. For instance, if one of the remaining answer choices is a direct opposite of one of the eliminated answer choices, it may well be the correct answer. The opposite of obviously wrong is obviously right! Of course, this is not always the case. Some answers are obviously incorrect simply because they are irrelevant to the question being asked. Still, identifying and eliminating some incorrect answer choices is a good way to simplify a multiple-choice question.

4. Don't Overanalyze

Anxious test takers often overanalyze questions. When you are nervous, your brain will often run wild causing you to make associations and discover clues that don't actually exist. If you feel that this may be a problem for you, do whatever you can to slow down during the test. Try taking a deep breath or counting to ten. As you read and consider the question, restrict yourself to the particular words used by the author. Avoid thought tangents about what the author *really* meant, or what he or she was *trying* to say. The only things that matter on a multiple-choice test are the words that are actually in the question. You must avoid reading too much into a multiple-choice question, or supposing that the writer meant something other than what he or she wrote.

5. No Need for Panic

It is wise to learn as many strategies as possible before taking a multiple-choice test, but it is likely that you will come across a few questions for which you simply don't know the answer. In this situation, avoid panicking. Because most multiple-choice tests include dozens of questions, the relative value of a single wrong answer is small. Moreover, your failure on one question has no effect on your success elsewhere on the test. As much as possible, you should compartmentalize each question on a multiple-choice test. In other words, you should not allow your feelings about one question to affect your success on the others. When you find a question that you either don't understand or don't know how to answer, just take a deep breath and do your best. Read the entire question slowly and carefully. Try rephrasing the question a couple of different ways. Then, read all of the answer choices carefully. After eliminating obviously wrong answers, make a selection and move on to the next question.

6. Confusing Answer Choices

When working on a difficult multiple-choice question, there may be a tendency to focus on the answer choices that are the easiest to understand. Many people, whether consciously or not, gravitate to the answer choices that require the least concentration, knowledge, and memory. This is a mistake. When you come across an answer choice that is confusing, you need to give it extra attention. A question might be confusing because you do not know the subject matter to which it refers. If this is the case, don't eliminate the answer before you have affirmatively settled on another. When you come across an answer choice of this type, set it aside as you look at the remaining choices. If you can confidently assert that one of the other choices is correct, you can leave the confusing answer aside. Otherwise, you will need to take a moment to try to better understand the confusing answer choice. Rephrasing is one way to tease out the sense of a confusing answer choice.

7. Your First Instinct

Many people struggle with multiple-choice tests because they overthink the questions. If you have studied sufficiently for the test, you should be prepared to trust your first instinct once you have carefully and completely read the question and all of the answer choices. There is a great deal of research suggesting that the mind can come to the correct conclusion very quickly once it has obtained all of the relevant information. At times, it may seem to you as if your intuition is working faster even than your reasoning mind. This may in fact be true. The knowledge you obtain while studying may be retrieved from your subconscious before you have a chance to work out the associations that support it. Verify your instinct by working out the reasons that it should be trusted.

8. Key Words

Many test takers struggle with multiple-choice questions because they have poor reading comprehension skills. Quickly reading and understanding a multiple-choice question requires a mixture of skill and experience. To help with this, try jotting down a few key words and phrases on a piece of scrap paper. Doing this concentrates the process of reading and forces the mind to weigh the relative importance of the question's parts. In selecting words and phrases to write down, the test taker thinks about the question more deeply and carefully. This is especially true for multiple-choice questions that are preceded by a long prompt.

9. Subtle Negatives

One of the oldest tricks in the multiple-choice test writer's book is to subtly reverse the meaning of a question with a word like *not* or *except*. If you are not paying attention to each word in the question, you can easily be led astray by this trick. For instance, a common question format is, "Which of the following is...?" Obviously, if the question instead is, "Which of the following is not....?," then the answer will be quite different. Even worse, the test makers are aware of the potential for this mistake and will include one answer choice that would be correct if the question were not negated or reversed. A test taker who misses the reversal will find what he or she believes to be a correct answer and will be so confident that he or she will fail to reread the question and discover the original error. The only way to avoid this is to practice a wide variety of multiple-choice questions and to pay close attention to each and every word.

10. Reading Every Answer Choice

It may seem obvious, but you should always read every one of the answer choices! Too many test takers fall into the habit of scanning the question and assuming that they understand the question because they recognize a few key words. From there, they pick the first answer choice that answers the question they believe they have read. Test takers who read all of the answer choices might discover that one of the latter answer choices is actually *more* correct. Moreover, reading all of the answer choices can remind you of facts related to the question that can help you arrive at the correct answer. Sometimes, a misstatement or incorrect detail in one of the latter answer choices will trigger your memory of the subject and will enable you to find the right answer. Failing to read all of the answer choices is like not reading all of the items on a restaurant menu: you might miss out on the perfect choice.

11. Spot the Hedges

One of the keys to success on multiple-choice tests is paying close attention to every word. This is never truer than with words like almost, most, some, and sometimes. These words are called "hedges", because they indicate that a statement is not totally true or not true in every place and time. An absolute statement will contain no hedges, but in many subjects, like literature and history, the answers are not always straightforward or absolute. There are always exceptions to the rules in these subjects. For this reason, you should favor those multiple-choice questions that contain hedging language. The presence of qualifying words indicates that the author is taking special care with his or her words, which is certainly important when composing the right answer. After all, there are many ways to be wrong, but there is only one way to be right! For this reason, it is wise to avoid answers that are absolute when taking a multiple-choice test. An absolute answer is one that says things are either all one way or all another. They often include words like *every*, *always*, *best*, and *never*. If you are taking a multiple-choice test in a subject that doesn't lend itself to absolute answers, be on your guard if you see any of these words.

12. Long Answers

In many subject areas, the answers are not simple. As already mentioned, the right answer often requires hedges. Another common feature of the answers to a complex or subjective question are qualifying clauses, which are groups of words that subtly modify the meaning of the sentence. If the question or answer choice describes a rule to which there are exceptions or the subject matter is complicated, ambiguous, or confusing, the correct answer will require many words in order to be expressed clearly and accurately. In essence, you should not be deterred by answer choices that seem excessively long. Oftentimes, the author of the text will not be able to write the correct answer without offering some qualifications and modifications. Your job is to read the answer choices thoroughly and completely and to select the one that most accurately and precisely answers the question.

13. Restating to Understand

Sometimes, a question on a multiple-choice test is difficult not because of what it asks but because of how it is written. If this is the case, restate the question or answer choice in different words. This process serves a couple of important purposes. First, it forces you to concentrate on the core of the question. In order to rephrase the question accurately, you have to understand it well. Rephrasing the question will concentrate your mind on the key words and ideas. Second, it will present the information to your mind in a fresh way. This process may trigger your memory and render some useful scrap of information picked up while studying.

14. True Statements

Sometimes an answer choice will be true in itself, but it does not answer the question. This is one of the main reasons why it is essential to read the question carefully and completely before proceeding to the answer choices. Too often, test takers skip ahead to the answer choices and look for true statements. Having found one of these, they are content to select it without reference to the question above. Obviously, this provides an easy way for test makers to play tricks. The savvy test taker will always read the entire question before turning to the answer choices. Then, having settled on a correct answer choice, he or she will refer to the original question and ensure that the selected answer is relevant. The mistake of choosing a correct-but-irrelevant answer choice is especially common on questions related to specific pieces of objective knowledge, like historical or scientific facts. A prepared test taker will have a wealth of factual knowledge at his or her disposal, and should not be careless in its application.

15. No Patterns

One of the more dangerous ideas that circulates about multiple-choice tests is that the correct answers tend to fall into patterns. These erroneous ideas range from a belief that B and C are the most common right answers, to the idea that an unprepared test-taker should answer "A-B-A-C-A-D-A-B-A." It cannot be emphasized enough that pattern-seeking of this type is exactly the WRONG way to approach a multiple-choice test. To begin with, it is highly unlikely that the test maker will plot the correct answers according to some predetermined pattern. The questions are scrambled and delivered in a random order. Furthermore, even if the test maker was following a pattern in the assignation of correct answers, there is no reason why the test taker would know which pattern he or she was using. Any attempt to discern a pattern in the answer choices is a waste of time and a distraction from the real work of taking the test. A test taker would be much better served by extra preparation before the test than by reliance on a pattern in the answers.

FREE DVD OFFER

Don't forget that doing well on your exam includes both understanding the test content and understanding how to use what you know to do well on the test. We offer a completely FREE Test Taking Tips DVD that covers world class test taking tips that you can use to be even more successful when you are taking your test.

All that we ask is that you email us your feedback about your study guide. To get your **FREE Test Taking Tips DVD**, email freedvd@studyguideteam.com with "FREE DVD" in the subject line and the following information in the body of the email:

- The title of your study guide.
- Your product rating on a scale of 1-5, with 5 being the highest rating.
- Your feedback about the study guide. What did you think of it?
- Your full name and shipping address to send your free DVD.

Introduction to the TSI Exam

Function of the Test

The Texas Success Initiative Assessment (TSI) is used to measure a student's readiness for college-level coursework. The TSI assesses incoming freshman students in the state of Texas in the subject areas of reading, writing and mathematics. The results of this test are used to place an incoming student into the appropriate college-level course and determine the type of intervention, if any, a student needs.

Many students can be exempt from the TSI if they met the minimum score on the SAT, ACT or statewide high school test. If students have already completed college-level English and math courses, they will be exempt. Other ways to be exempt from taking the TSI include current enrollment in a Level-One certificate program (fewer than 43 semester credit hours), not seeking a college degree, or students currently or formerly enrolled in the military. Those students that are not exempt will be asked to take three tests, one in each of the subject areas of mathematics, reading, and writing. A diagnostic test may be given in addition to these three tests in a particular subject area if deemed necessary.

The TSI is meant to give an institute detailed information on a student's strengths and weaknesses. Before taking the TSI, a candidate must take the Pre-Assessment activity. The activity includes an explanation of the TSI, practice questions and feedback, an explanation of developmental educational options if a minimum score is not met, and campus and community resources.

Test Administration

There are testing centers located at universities for the TSI assessment. Retesting is acceptable, and there are no restrictions as to when a retest can be done. However, students are encouraged to study and prepare before retaking the TSI. Some colleges and universities also host workshops to better prepare a student to retake the TSI. It is advised to contact the local university or college where a candidate wishes to take the test to receive more information on retesting and workshops. Advisors and counselors at a particular institution can better help a candidate with any further questions they may have.

Test Format

The TSI has three different components: one in mathematics, one in reading and one in writing. All sections are multiple choice with an additional essay section for writing. The table below outlines the concepts tested in each of the sections.

Mathematics	**Reading**	**Writing**
Elementary Algebra and Functions	Literary Analysis	Essay Revision
Intermediate Algebra and Functions	Main Idea and Supporting Details	Agreement
Geometry and Measurement	Inferences in a Text or Texts	Sentence Structure
Data Analysis, Statistics, and Probability	Author's Use of Language	Sentence Logic

The essay section is a five paragraph persuasive essay on a controversial issue or current event topic. The essay should be 300-600 words and must clearly state a main idea with specific evidence to support the main idea. Correct conventions are expected, and no dictionaries or resources may be used. Scratch paper and essay paper will be provided at the testing center.

Section	Approximate Number of Items on Placement Test	Approximate Number of Items on Diagnostic Test
Mathematics	20	10
Reading	24	10-12
Writing	20	10-12

Scoring

The minimum passing score for the TSI Assessment to determine a student's readiness for freshman college-level coursework is different for each subject area. A TSI score of 351 or above is necessary for reading. A TSI score of at least 350 and an essay score of at least 5 is required for writing. Other ways to meet requirements for writing would be a score of at least 363 and an essay score of 4 or a score with less than 350 but with an ABE diagnostic level of at least 4 and an essay score of at least 5. A TSI score of 350 or above is required for math.

Recent/Future Developments

The minimum passing TSI standards and scores noted above in the scoring section are subject to change in Fall 2017 on the first day of classes. The mathematics and reading standards are the ones most likely to change.

Mathematics

Elementary Algebra and Functions

Relations and Functions

First, it's important to understand the definition of a *relation*. Given two variables, *x* and *y*, which stand for unknown numbers, a *relation* between *x* and *y* is an object that splits all of the pairs (*x*, *y*) into those for which the relation is true and those for which it is false. For example, consider the relation of $x^2 = y^2$. This relationship is true for the pair (1, 1) and for the pair (-2, 2), but false for (2, 3). Another example of a relation is $x \leq y$. This is true whenever *x* is less than or equal to *y*.

A *function* is a special kind of relation where, for each value of *x*, there is only a single value of *y* that satisfies the relation. So, $x^2 = y^2$ is *not* a function because in this case, if *x* is 1, *y* can be either 1 or -1: the pair (1, 1) and (1, -1) both satisfy the relation. More generally, for this relation, any pair of the form $(a, \pm a)$ will satisfy it. On the other hand, consider the following relation: $y = x^2 + 1$. This is a function because for each value of *x*, there is a unique value of *y* that satisfies the relation. Notice, however, there are multiple values of *x* that give us the same value of *y*. This is perfectly acceptable for a function. Therefore, *y* is a function of *x*.

To determine if a relation is a function, check to see if every *x* value has a unique corresponding *y* value.

A function can be viewed as an object that has *x* as its input and outputs a unique *y*-value. It is sometimes convenient to express this using *function notation*, where the function itself is given a name, often *f*. To emphasize that *f* takes *x* as its input, the function is written as $f(x)$. In the above example, the equation could be rewritten as $f(x) = x^2 + 1$. To write the value that a function yields for some specific value of *x*, that value is put in place of *x* in the function notation. For example, $f(3)$ means the value that the function outputs when the input value is 3. If $f(x) = x^2 + 1$, then $f(3) = 3^2 + 1 = 10$.

A function can also be viewed as a table of pairs (*x*, *y*), which lists the value for *y* for each possible value of *x*.

The set of all possible values for *x* in $f(x)$ is called the *domain* of the function, and the set of all possible outputs is called the *range* of the function. Note that usually the domain is assumed to be all real numbers, except those for which the expression for $f(x)$ is not defined, unless the problem specifies otherwise. An example of how a function might not be defined is in the case of $f(x) = \frac{1}{x+1}$, which is not defined when $x = -1$ (which would require dividing by zero). Therefore, in this case the domain would be all real numbers except $x = -1$.

If *y* is a function of *x*, then *x* is the *independent variable* and *y* is the *dependent variable*. This is because in many cases, the problem will start with some value of *x* and then see how *y* changes depending on this starting value.

Evaluating Functions

To evaluate functions, plug in the given value everywhere the variable appears in the expression for the function. For example, find $g(-2)$ where $g(x) = 2x^2 - \frac{4}{x}$. To complete the problem, plug in -2 in the following way: $g(-2) = 2(-2)^2 - \frac{4}{-2} = 2 \times 4 + 2 = 8 + 2 = 10$.

Defining Linear Equations

A function is called *linear* if it can take the form of the equation $f(x) = ax + b$, or $y = ax + b$, for any two numbers a and b. A linear equation forms a straight line when graphed on the coordinate plane. An example of a linear function is shown below on the graph.

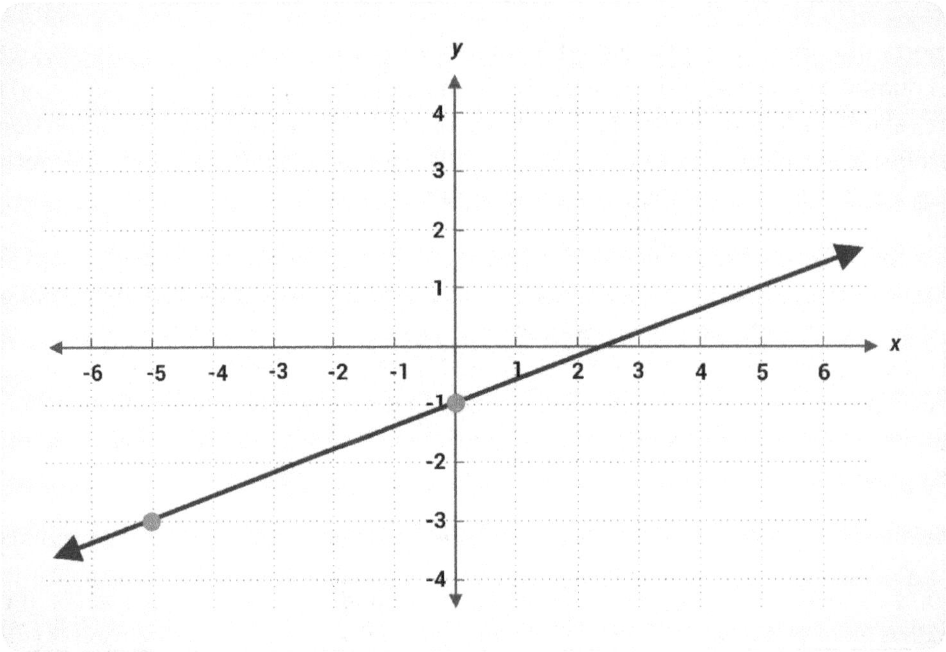

This is a graph of the following function: $y = \frac{2}{5}x - 1$. A table of values that satisfies this function is shown below.

x	y
-5	-3
0	-1
5	1
10	3

These points can be found on the graph using the form (x,y). For more on graphing in the coordinate plane, refer to the Graphing section below.

Graphing Functions and Relations

To graph relations and functions, the Cartesian plane is used. This means to think of the plane as being given a grid of squares, with one direction being the x-axis and the other direction the y-axis. Generally, the independent variable is placed along the horizontal axis, and the dependent variable is placed along the vertical axis. Any point on the plane can be specified by saying how far to go along the x-axis and how far along the y-axis with a pair of numbers (x, y). Specific values for these pairs can be given names such as $C = (-1, 3)$. Negative values mean to move left or down; positive values mean to move right or up. The point where the axes cross one another is called the *origin*. The origin has coordinates $(0, 0)$ and is usually called O when given a specific label. An illustration of the Cartesian plane, along with graphs of $(2, 1)$ and $(-1, -1)$, are below.

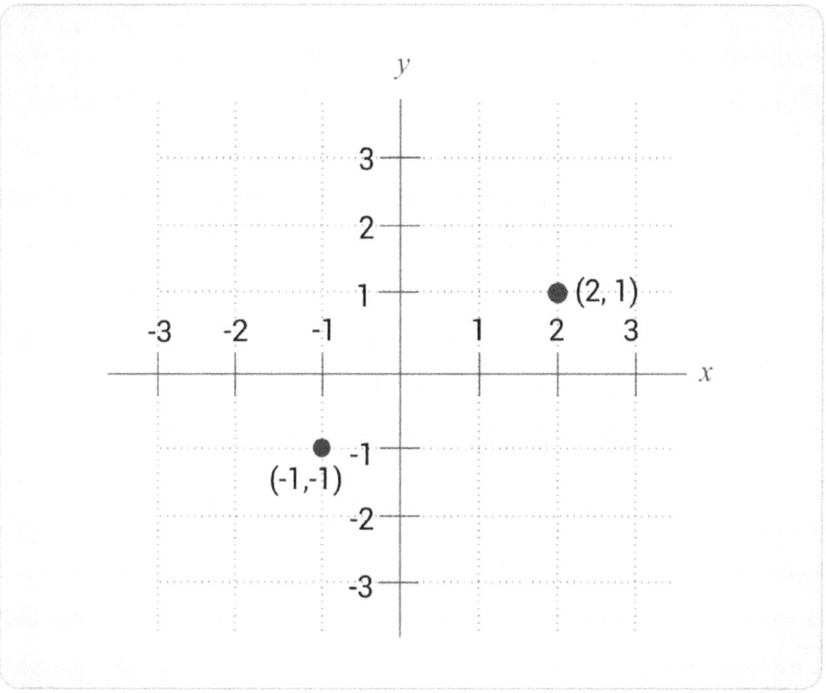

Relations also can be graphed by marking each point whose coordinates satisfy the relation. If the relation is a function, then there is only one value of y for any given value of x. This leads to the **vertical line test**: if a relation is graphed, then the relation is a function if any possible vertical line drawn anywhere along the graph would only touch the graph of the relation in no more than one place. Conversely, when graphing a function, then any possible vertical line drawn will not touch the graph of the function at any point or will touch the function at just one point. This test is made from the definition of a function, where each x-value must be mapped to one and only one y-value.

Forms of Linear Equations

When graphing a linear function, note that the ratio of the change of the y coordinate to the change in the x coordinate is constant between any two points on the resulting line, no matter which two points are chosen. In other words, in a pair of points on a line, (x_1, y_1) and (x_2, y_2), with $x_1 \neq x_2$ so that the two points are distinct, then the ratio $\frac{y_2 - y_1}{x_2 - x_1}$ will be the same, regardless of which particular pair of points are chosen. This ratio, $\frac{y_2-y_1}{x_2-x_1}$, is called the *slope* of the line and is frequently denoted with the letter m. If slope m is positive, then the line goes upward when moving to the right, while if slope m is negative, then the line goes downward when moving to the right. If the slope is 0, then the line is called *horizontal*, and the *y* coordinate is constant along the entire line. In lines where the *x* coordinate is constant along the entire line, *y* is not actually a function of *x*. For such lines, the slope is not defined. These lines are called *vertical* lines.

Linear functions may take forms other than $y = ax + b$. The most common forms of linear equations are explained below:

1. Standard Form: $Ax + By = C$, in which the slope is given by $m = \frac{-A}{B}$, and the *y*-intercept is given by $\frac{C}{B}$.
2. Slope-Intercept Form: $y = mx + b$, where the slope is *m* and the *y* intercept is *b*.
3. Point-Slope Form: $y - y_1 = m(x - x_1)$, where the slope is *m* and (x_1, y_1) is any point on the chosen line.
4. Two-Point Form: $\frac{y-y_1}{x-x_1} = \frac{y_2-y_1}{x_2-x_1}$, where (x_1, y_1) and (x_2, y_2) are any two distinct points on the chosen line. Note that the slope is given by $m = \frac{y_2-y_1}{x_2-x_1}$.
5. Intercept Form: $\frac{x}{x_1} + \frac{y}{y_1} = 1$, in which x_1 is the *x*-intercept and y_1 is the *y*-intercept.

These five ways to write linear equations are all useful in different circumstances. Depending on the given information, it may be easier to write one of the forms over another.

If $y = mx$, *y* is directly proportional to *x*. In this case, changing *x* by a factor changes *y* by that same factor. If $y = \frac{m}{x}$, *y* is inversely proportional to *x*. For example, if *x* is increased by a factor of 3, then *y* will be decreased by the same factor, 3.

Solving Linear Equations

Sometimes, rather than a situation where there's an equation such as $y = ax + b$ and finding *y* for some value of *x* is requested, the result is given and finding *x* is requested.

The key to solving any equation is to remember that from one true equation, another true equation can be found by adding, subtracting, multiplying, or dividing both sides by the same quantity. In this case, it's necessary to manipulate the equation so that one side only contains *x*. Then the other side will show what *x* is equal to.

For example, in solving $3x - 5 = 2$, adding 5 to each side results in $3x = 7$. Next, dividing both sides by 3 results in $x = \frac{7}{3}$. To ensure the value of x is correct, the number can be substituted into the original equation and solved to see if it makes a true statement. For example, $3(\frac{7}{3}) - 5 = 2$ can be simplified by cancelling out the two 3s. This yields $7 - 5 = 2$, which is a true statement.

Sometimes an equation may have more than one *x* term. For example, consider the following equation: $3x + 2 = x - 4$. Moving all of the *x* terms to one side by subtracting *x* from both sides results in $2x +$

$2 = -4$. Next, subtract 2 from both sides so that there is no constant term on the left side. This yields $2x = -6$. Finally, divide both sides by 2, which leaves $x = -3$.

Solving Linear Inequalities

Solving linear inequalities is very similar to solving equations, except for one rule: when multiplying or dividing an inequality by a negative number, the inequality symbol changes direction. Given the following inequality, solve for x: $-2x + 5 < 13$. The first step in solving this equation is to subtract 5 from both sides. This leaves the inequality: $-2x < 8$. The last step is to divide both sides by -2. By using the rule, the answer to the inequality is $x > -4$.

Since solutions to inequalities include more than one value, number lines are used many times to model the answer. For the previous example, the answer is modelled on the number line below. It shows that any number greater than -4, not including -4, satisfies the inequality.

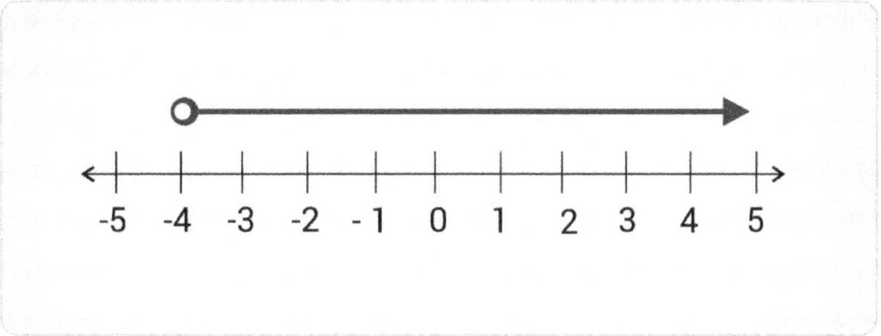

Linear Systems of Equations

A problem sometimes involves multiple variables and multiple equations. These are called *systems of equations*. In this case, try to manipulate them until an expression is found that provides the value of one of the variables. There are a couple of different approaches to this, and some of them can be used together in some cases. The three basic rules to keep in mind are the following.

1. Manipulate a set of equations by doing the same operation to both equations, just as is done when working with just one equation.

2. If one of the equations can be changed so that it expresses one variable in terms of the others, then that expression can be substituted into the other equations and the variable can be eliminated. This means the other equations will have one less variable in them. This is called the method of substitution.

3. If two equations of the form $a = b, c = d$ are included, then a new equation can be formed by adding the left sides and adding the right sides, $a + c = b + d$, or $a - c = b - d$. This enables the elimination of one of the variables from an equation. This is called the method of elimination.

The simplest case is the case of a *linear* system of equations. Although the equations may be written in more complicated forms, linear systems of equations with two variables can always be written in the form $ax + by = c, dx + ey = f$. The two basic approaches to solving these systems are substitution and elimination.

Consider the system $3x - y = 2 \text{ and } 2x + 2y = 3$. This can be solved in two ways.

1. By substitution: start by solving the first equation for y. First, subtract $3x$ from both sides to obtain $-y = 2 - 3x$. Next, divide both sides by -1, to obtain $y = 3x - 2$. Then substitute this value for y into the second equation. This yields $2x + 2(3x - 2) = 3$. This can be simplified to $2x + 6x - 4 = 3$, or $8x = 7$, which means $x = \frac{7}{8}$. By plugging in this value for x into $y = 3x - 2$, the result is $y = 3\left(\frac{7}{8}\right) - 2 = \frac{21}{8} - \frac{16}{8} = \frac{5}{8}$. So, this results in $x = \frac{7}{8}, y = \frac{5}{8}$.

2. By elimination: first, multiply the first equation by 2. This results in $-2y$, which could cancel out the $+2y$ in the second equation. Multiplying both sides of the first equation by 2 gives results in $2(3x - y) = 2(2)$, or $6x - 2y = 4$. Adding the left sides and the right sides of the two equations and setting the results equal to one another results in $(6x + 2x) + (-2y + 2y) = 4 + 3$. This simplifies to $8x = 7$, so again $x = \frac{7}{8}$. Plug this back into either of the original equations and the result is $3\left(\frac{7}{8}\right) - y = 2$ or $y = 3\left(\frac{7}{8}\right) - 2 = \frac{21}{8} - \frac{16}{8} = \frac{5}{8}$. This again yields $x = \frac{7}{8}, y = \frac{5}{8}$.

As this shows, both methods will give the same answer. However, one method is sometimes preferred over another simply because of the amount of work required. To check the answer, the values can be substituted into the given system to make sure they form two true statements.

Algebraic Expressions and Equations

Algebraic expressions look similar to equations, but they do not include the equal sign. Algebraic expressions are comprised of numbers, variables, and mathematical operations. Some examples of algebraic expressions are $8x + 7y - 12z$, $3a^2$, and $5x^3 - 4y^4$.

Algebraic expressions and equations can be used to represent real-life situations and model the behavior of different variables. For example, $2x + 5$ could represent the cost to play games at an arcade. In this case, 5 represents the price of admission to the arcade and 2 represents the cost of each game played. To calculate the total cost, use the number of games played for x, multiply it by 2, and add 5.

Word Problems and Applications

In word problems, multiple quantities are often provided with a request to find some kind of relation between them. This often will mean that one variable (the dependent variable whose value needs to be found) can be written as a function of another variable (the independent variable whose value can be figured from the given information). The usual procedure for solving these problems is to start by giving each quantity in the problem a variable, and then figuring the relationship between these variables.

For example, suppose a car gets 25 miles per gallon. How far will the car travel if it uses 2.4 gallons of fuel? In this case, *y* would be the distance the car has traveled in miles, and *x* would be the amount of fuel burned in gallons (2.4). Then the relationship between these variables can be written as an algebraic equation, $y = 25x$. In this case, the equation is $y = 25 \times 2.4 = 60$, so the car has traveled 60 miles.

Some word problems require more than just one simple equation to be written and solved. Consider the following situations and the linear equations used to model them.

Suppose Margaret is 2 miles to the east of John at noon. Margaret walks to the east at 3 miles per hour. How far apart will they be at 3 p.m.? To solve this, *x* would represent the time in hours past noon, and *y* would represent the distance between Margaret and John. Now, noon corresponds to the equation where *x* is 0, so the *y* intercept is going to be 2. It's also known that the slope will be the rate at which the distance is changing, which is 3 miles per hour. This means that the slope will be 3 (be careful at this point: if units were used, other than miles and hours, for *x* and *y* variables, a conversion of the given information to the appropriate units would be required first). The simplest way to write an equation given the *y*-intercept, and the slope is the Slope-Intercept form, is $y = mx + b$. Recall that *m* here is the slope and *b* is the *y* intercept. So, $m = 3$ and $b = 2$. Therefore, the equation will be $y = 3x + 2$. The word problem asks how far to the east Margaret will be from John at 3 p.m., which means when *x* is 3. So, substitute $x = 3$ into this equation to obtain $y = 3 \times 3 + 2 = 9 + 2 = 11$. Therefore, she will be 11 miles to the east of him at 3 p.m.

For another example, suppose that a box with 4 cans in it weighs 6 lbs., while a box with 8 cans in it weighs 12 lbs. Find out how much a single can weighs. To do this, let *x* denote the number of cans in the box, and *y* denote the weight of the box with the cans in lbs. This line touches two pairs: $(4, 6)$ and (8, 12). A formula for this relation could be written using the two-point form, with $x_1 = 4, y_1 = 6, x_2 = 8, y_2 = 12$. This would yield $\frac{y-6}{x-4} = \frac{12-6}{8-4}$, or $\frac{y-6}{x-4} = \frac{6}{4} = \frac{3}{2}$. However, only the slope is needed to solve this problem, since the slope will be the weight of a single can. From the computation, the slope is $\frac{3}{2}$. Therefore, each can weighs $\frac{3}{2}$ lb.

Intermediate Algebra and Functions

Polynomials

An expression of the form ax^n, where *n* is a non-negative integer, is called a *monomial* because it contains one term. A sum of monomials is called a *polynomial*. For example, $-4x^3 + x$ is a polynomial, while $5x^7$ is a monomial. A function equal to a polynomial is called a *polynomial function*.

The monomials in a polynomial are also called the *terms* of the polynomial.

The constants that precede the variables are called *coefficients*.

The highest value of the exponent of *x* in a polynomial is called the *degree* of the polynomial. So, $-4x^3 + x$ has a degree of 3, while $-2x^5 + x^3 + 4x + 1$ has a degree of 5. When multiplying polynomials, the degree of the result will be the sum of the degrees of the two polynomials being multiplied.

To add polynomials, add the coefficients of like powers of *x*. For example, $(-2x^5 + x^3 + 4x + 1) + (-4x^3 + x) = -2x^5 + (1 - 4)x^3 + (4 + 1)x + 1 = -2x^5 - 3x^3 + 5x + 1$.

Likewise, subtraction of polynomials is performed by subtracting coefficients of like powers of *x*. So, $(-2x^5 + x^3 + 4x + 1) - (-4x^3 + x) = -2x^5 + (1 + 4)x^3 + (4 - 1)x + 1 = -2x^5 + 5x^3 + 3x + 1$.

To multiply two polynomials, multiply each term of the first polynomial by each term of the second polynomial and add the results. For example, $(4x^2 + x)(-x^3 + x) = 4x^2(-x^3) + 4x^2(x) + x(-x^3) + x(x) = -4x^5 + 4x^3 - x^4 + x^2$. In the case where each polynomial has two terms, like in this example, some students find it helpful to remember this as multiplying the First terms, then the Outer terms, then the Inner terms, and finally the Last terms, with the mnemonic FOIL. For longer polynomials, the multiplication process is the same, but there will be, of course, more terms, and there is no common mnemonic to remember each combination.

The process of *factoring* a polynomial means to write the polynomial as a product of other (generally simpler) polynomials. Here is an example: $x^2 - 4x + 3 = (x - 1)(x - 3)$. If a certain monomial divides every term of the polynomial, factor it out of each term, for example:

$$4x^3 + 16x^2 = 4x^2(x + 4).$$

$$x^2 + 2xy + y^2 = (x + y)^2 \text{ or } x^2 - 2xy + y^2 = (x - y)^2$$

$$x^2 - y^2 = (x + y)(x - y)$$

$$x^3 + y^3 = (x + y)(x^2 - xy + y^2)$$

$$x^3 - y^3 = (x - y)(x^2 + xy + y^2)$$

$$x^3 + 3x^2y + 3xy^2 + y^3 = (x + y)^3 \text{ and } x^3 - 3x^2y + 3xy^2 - y^3 = (x - y)^3$$

It sometimes can be necessary to rewrite the polynomial in some clever way before applying the above rules. Consider the problem of factoring $x^4 - 1$. This does not immediately look like any of the cases for which there are rules. However, it's possible to think of this polynomial as $x^4 - 1 = (x^2)^2 - (1^2)^2$, and now apply the third rule in the above list to simplify this: $(x^2)^2 - (1^2)^2 = (x^2 + 1^2)(x^2 - 1^2) = (x^2 + 1)(x^2 - 1)$.

Quadratic Functions

A polynomial of degree 2 is called *quadratic*. Every quadratic function can be written in the form $ax^2 + bx + c$. The graph of a quadratic function, $y = ax^2 + bx + c$, is called a *parabola*. Parabolas are vaguely U-shaped.

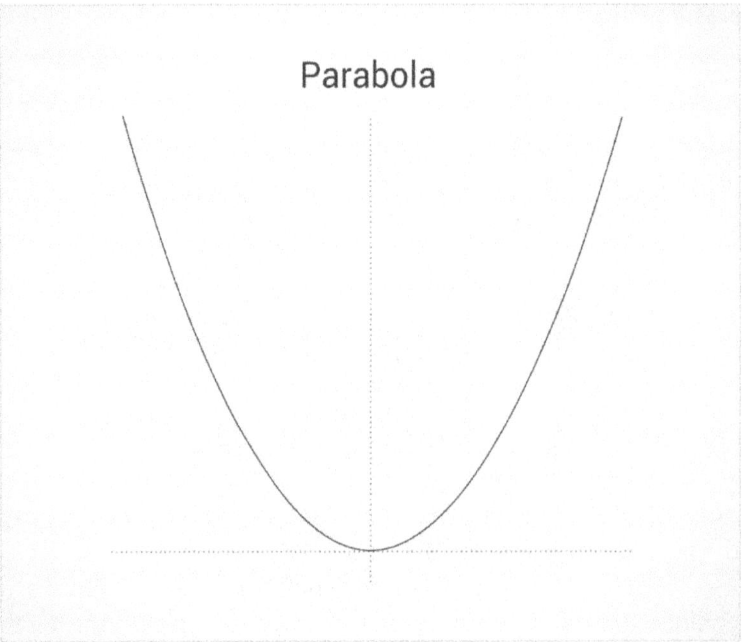

Whether the parabola opens upward or downward depends on the sign of a. If a is positive, then the parabola will open upward. If a is negative, then the parabola will open downward. The value of a will also affect how wide the parabola is. If the absolute value of a is large, then the parabola will be fairly skinny. If the absolute value of a is small, then the parabola will be quite wide.

Changes to the value of b affect the parabola in different ways, depending on the sign of a. For positive values of a, increasing b will move the parabola to the left, and decreasing b will move the parabola to the right. On the other hand, if a is negative, the effects will be the opposite: increasing b will move the parabola to the right, while decreasing b will move the parabola to the left.

Changes to the value of c move the parabola vertically. The larger that c is, the higher the parabola gets. This does not depend on the value of a.

The quantity $D = b^2 - 4ac$ is called the *discriminant* of the parabola. When the discriminant is positive, then the parabola has two real zeros, or x intercepts. However, if the discriminant is negative, then there are no real zeros, and the parabola will not cross the x-axis. The highest or lowest point of the parabola is called the *vertex*. If the discriminant is zero, then the parabola's highest or lowest point is on the x-axis, and it will have a single real zero. The x-coordinate of the vertex can be found using the equation $x = -\frac{b}{2a}$. Plug this x-value into the equation and find the y-coordinate.

A quadratic equation is often used to model the path of an object thrown into the air. The x-value can represent the time in the air, while the y-value can represent the height of the object. In this case, the maximum height of the object would be the y-value found when the x-value is $-\frac{b}{2a}$.

Solving Quadratic Equations

A *quadratic equation* is an equation in the form $ax^2 + bx + c = 0$. There are several methods to solve such equations. The easiest method will depend on the quadratic equation in question.

It sometimes is possible to solve quadratic equations by manually *factoring* them. This means rewriting them in the form $(x + A)(x + B) = 0$. If this is done, then they can be solved by remembering that when $ab = 0$, either a or b must be equal to zero. Therefore, to have $(x + A)(x + B) = 0$, $(x + A) = 0$ or $(x + B) = 0$ is needed. These equations have the solutions $x = -A$ and $x = -B$, respectively.

In order to factor a quadratic equation, note that $(x + A)(x + B) = x^2 + (A + B)x + AB$. So, if an equation is in the form $x^2 + bx + c$, two numbers, A and B, need to be found that will add up to give us b, and multiply together to give us c.

As an example, consider solving the equation $-3x^2 + 6x + 9 = 0$. Start by dividing both sides by -3, leaving $x^2 - 2x - 3 = 0$. Now, notice that $1 - 3 = -2$, and also that $(1)(-3) = -3$. This means the equation can be factored into $(x + 1)(x - 3) = 0$. Now, solve $(x + 1) = 0$ and $(x - 3) = 0$ to get $x = -1$ and $x = 3$ as the solutions.

It is useful when trying to factor to remember that $x^2 + 2xy + y^2 = (x + y)^2$, $x^2 - 2xy + y^2 = (x - y)^2$, and $x^2 - y^2 = (x + y)(x - y)$.

However, factoring by hand is often hard to do. If there are no obvious ways to factor the quadratic equation, solutions can still be found by using the *quadratic formula*.

The quadratic formula is:

$$x = \frac{-b \pm \sqrt{b^2 - 4ac}}{2a}$$

This method will always work, although it sometimes can take longer than factoring by hand, if the factors are easy to guess. Using the standard form $ax^2 + bx + c = 0$, plug the values of *a*, *b*, and *c* from the equation into the formula and solve for x. There will either be two answers, one answer, or no real answer. No real answer comes when the value of the discriminant, the number under the square root, is a negative number. Since there are no real numbers that square to get a negative, the answer will be no real roots.

Here is an example of solving a quadratic equation using the quadratic formula. Suppose the equation to solve is $-2x^2 + 3x + 1 = 0$. There is no obvious way to factor this, so the quadratic formula is used, with $a = -2, b = 3, c = 1$. After substituting these values into the quadratic formula, it yields this:

$$x = \frac{-3 \pm \sqrt{3^2 - 4(-2)(1)}}{2(-2)}$$

This can be simplified to obtain:

$$\frac{3 \pm \sqrt{9 + 8}}{4}$$

or

$$\frac{3 \pm \sqrt{17}}{4}$$

Challenges can be encountered when asked to find a quadratic equation with specific roots. Given roots *A* and *B*, a quadratic function can be constructed with those roots by taking $(x - A)(x - B)$. So, in constructing a quadratic equation with roots $x = -2, 3$, it would result in $(x + 2)(x - 3) = x^2 - x - 6$. Multiplying this by a constant also could be done without changing the roots.

Exponents and Roots

An *exponent* is written as a^b. In this expression, *a* is called the *base* and *b* is called the *exponent*. It is properly stated that *a* is raised to the *n*-th power. Therefore, in the expression 2^3, the exponent is 3, while the base is 2. Such an expression is called an *exponential expression*. Note that when the exponent is 2, it is called *squaring* the base, and when it is 3, it is called *cubing* the base.

When the exponent is a positive integer, this indicates the base is multiplied by itself the number of times written in the exponent. So, in the expression 2^3, multiply 2 by itself with 3 copies of 2: $2^3 = 2 \times 2 \times 2 = 8$. One thing to notice is that, for positive integers *n* and *m*, $a^n a^m = a^{n+m}$ is a rule. In order to make this rule be true for an integer of 0, $a^0 = 1$, so that $a^n a^0 = a^{n+0} = a^n$. And, in order to make this rule be true for negative exponents, $a^{-n} = \frac{1}{a^n}$.

Another rule for simplifying expressions with exponents is shown by the following equation: $(a^m)^n = a^{mn}$. This is true for fractional exponents as well. So, for a positive integer, define $a^{\frac{1}{n}}$ to be the number that, when raised to the *n*-th power, provides *a*. In other words, $(a^{\frac{1}{n}})^n = a$ is the desired equation. It should be noted that $a^{\frac{1}{n}}$ is the *n*-th root of *a*. This also can be written as $a^{\frac{1}{n}} = \sqrt[n]{a}$. The symbol on the right-hand side of this equation is called a *radical*. If the root is left out, assume that the 2nd root should be taken, also called the *square* root: $a^{\frac{1}{2}} = \sqrt[2]{a} = \sqrt{a}$. Additionally, $\sqrt[3]{a}$ is also called the *cube* root.

Note that when multiple roots exist, $a^{\frac{1}{n}}$ is defined to be the *positive* root. So, $4^{\frac{1}{2}} = 2$. Also note that negative numbers do not have even roots in the real numbers.

This also enables finding exponents for any rational number: $a^{\frac{m}{n}} = (a^{\frac{1}{n}})^m = (a^m)^{\frac{1}{n}}$. In fact, the exponent can be any real number. In general, the following rules for exponents should be used for any numbers $a, b, m,$ and n.

- $a^1 = a.$
- $1^a = 1.$
- $a^0 = 1.$
- $a^m a^n = a^{m+n}.$
- $\frac{a^m}{a^n} = a^{m-n}$
- $(a^m)^n = a^{m \times n}.$
- $(ab)^m = a^m b^m.$
- $(\frac{a}{b})^m = \frac{a^m}{b^m}.$

As an example of applying these rules, consider the problem of simplifying the expression $(3x^2y)^3(2xy^4)$. Start by simplifying the left term using the sixth rule listed. Applying this rule yields the following expression: $27x^6y^3(2xy^4)$. The exponents can now be combined with base *x* and the exponents with base *y*. Multiply the coefficients to yield $54x^7y^7$.

Order of Operations

When working with complicated expressions, parentheses are used to indicate in which order to perform operations. However, to avoid having too many parentheses in an expression, here are some basic rules concerning the proper order to perform operations when not otherwise specified.

1. Parentheses: always perform operations inside parentheses first, regardless of what those operations are
2. Exponents
3. Multiplication and Division
4. Addition and Subtraction

For #3 & #4, work these from left to right. So, if there a subtraction problem and then an addition problem, the subtraction problem will be worked first.

Note that multiplication and division are performed from left to right as they appear in the expression or equation. Addition and subtraction also are performed from left to right as they appear.

As an aid to memorizing this, some students like to use the mnemonic PEMDAS. Furthermore, this acronym can be associated with a mnemonic phrase such as "Pirates Eat Many Donuts At Sea."

Solving Equations with Exponents and Roots

Here are some of the most important properties of exponents and roots: if *n* is an integer, and if $a^n = b^n$, then $a = b$ if *n* is odd; but $a = \pm b$ if *n* is even. Similarly, if the roots of two things are equal, $\sqrt[n]{a} = \sqrt[n]{b}$, then $a = b$. This means that when starting with a true equation, both sides of that equation can be raised to a given power to obtain another true equation. Beware that when an even-powered root is taken on both sides of the equation, a $\pm$ in the result. For example, given the equation $x^2 = 16$, take the square root of both sides to solve for x. This results in the answer $x = \pm 4$ because $(-4)^2 = 16$ and $(4)^2 = 16$.

Another property is that if $a^n = a^m$, then $n = m$. This is true for any real numbers *n* and *m*.

For solving the equation $\sqrt{x+2} - 1 = 3$, start by moving the -1 over to the right-hand side. This is performed by adding 1 to both sides, which yields $\sqrt{x+2} = 4$. Now, square both sides, but remember that by squaring both sides, the signs are irrelevant. This yields $x + 2 = 16$, which simplifies to give $x = 14$.

Now consider the problem $(x+1)^4 = 16$. To solve this, take the 4th root of both sides, which means an ambiguity in the sign will be introduced because it is an even root: $\sqrt[4]{(x+1)^4} = \pm\sqrt[4]{16}$. The right-hand side is 2, since $2^4 = 16$. Therefore, $x + 1 = \pm 2$ or $x = -1 \pm 2$. Thus, the two possible solutions are $x = -3$ and $x = 1$.

Remember that when solving equations, the answer can be checked by plugging the solution back into the problem to make a true statement.

Rational Expressions

A rational expression is a fraction where the numerator and denominator are both polynomials. Some examples of rational expressions include the following: $\frac{4x^3y^5}{3z^4}$, $\frac{4x^3+3x}{x^2}$, and $\frac{x^2+7x+10}{x+2}$. Since these refer to expressions and not equations, they can be simplified but not solved. Using the rules in the previous Exponents and Roots sections, some rational expressions with monomials can be simplified. Other rational expressions such as the last example, $\frac{x^2+7x+10}{x+2}$, take more steps to be simplified. First, the polynomial on top can be factored from $x^2 + 7x + 10$ into $(x+5)(x+2)$. Then the common factors can be canceled and the expression can be simplified to $(x+5)$.

Consider this problem as an example of using rational expressions. Reggie wants to lay sod in his rectangular backyard. The length of the yard is given by the expression $4x + 2$ and the width is unknown. The area of the yard is $20x + 10$. Reggie needs to find the width of the yard. Knowing that the area of a rectangle is length multiplied by width, an expression can be written to find the width: $\frac{20x+10}{4x+2}$, area divided by length. Simplifying this expression by factoring out 10 on the top and 2 on the bottom leads to this expression: $\frac{10(2x+1)}{2(2x+1)}$. By cancelling out the $2x + 1$, that results in $\frac{10}{2} = 5$. The width of the yard is found to be 5 by simplifying a rational expression.

Rational Equations

A *rational equation* can be as simple as an equation with a ratio of polynomials, $\frac{p(x)}{q(x)}$, set equal to a value, where $p(x)$ and $q(x)$ are both polynomials. Notice that a rational equation has an equal sign, which is different from expressions. This leads to solutions, or numbers that make the equation true.

It is possible to solve rational equations by trying to get all of the *x* terms out of the denominator and then isolate them on one side of the equation. For example, to solve the equation $\frac{3x+2}{2x+3} = 4$, start by multiplying both sides by $(2x + 3)$. This will cancel on the left side to yield $3x + 2 = 4(2x + 3)$, then $3x + 2 = 8x + 12$. Now, subtract $8x$ from both sides, which yields $-5x + 2 = 12$. Subtracting 2 from both sides results in $-5x = 10$. Finally, divide both sides by -5 to obtain $x = -2$.

Sometimes, when solving rational equations, it can be easier to try to simplify the rational expression by factoring the numerator and denominator first, then cancelling out common factors. For example, to solve $\frac{2x^2-8x+6}{x^2-3x+2} = 1$, start by factoring $2x^2 - 8x + 6 = 2(x^2 - 4x + 3) = 2(x - 1)(x - 3)$. Then, factor $x^2 - 3x + 2$ into $(x - 1)(x - 2)$. This turns the original equation into $\frac{2(x-1)(x-3)}{(x-1)(x-2)} = 1$. The common factor of $(x - 1)$ can be canceled, leaving us with $\frac{2(x-3)}{x-2} = 1$. Now the same method used in the previous example can be followed. Multiplying both sides by $x - 2$ and performing the multiplication on the left yields $2x - 6 = x - 2$, which can be simplified to $x = 4$.

Rational Functions

A rational function is similar to an equation, but it includes two variables. In general, a rational function is in the form: $f(x) = \frac{p(x)}{q(x)}$, where $p(x)$ and $q(x)$ are polynomials. Refer to the previous Functions section for a more detailed definition of functions. Rational functions are defined everywhere except where the denominator is equal to zero. When the denominator is equal to zero, this indicates either a hole in the graph or an asymptote. An asymptote can be either vertical, horizontal, or slant. A hole occurs when both the numerator and denominator are equal to 0 for a given value of *x*.

A rational function can have at most one vertical asymptote and one horizontal or slant asymptote. An asymptote is a line such that the distance between the curve and the line tends toward 0, but never reaches it, as the line heads toward infinity. Examples of these types of functions are shown below. The first graph shows a rational function with a vertical asymptote at x = 0. This can be found by setting the denominator equal to 0. In this case it is just x = 0.

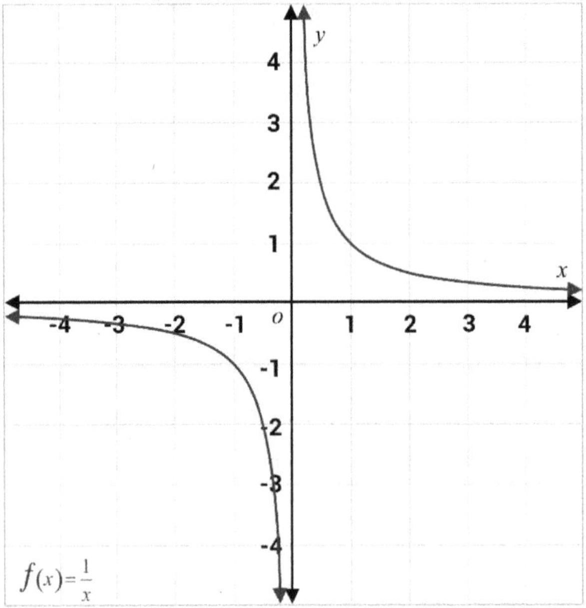

The second graph shows a rational function with a vertical asymptote at x = -.5. Again this can be found by just setting the denominator equal to 0. So, $2x^2 + x = 0, 2x + 1 = 0, 2x = -1, x = -.5$. This graph also has a hole in the graph at $x = 0$. This is because both the numerator and denominator are equal to 0 when $x = 0$.

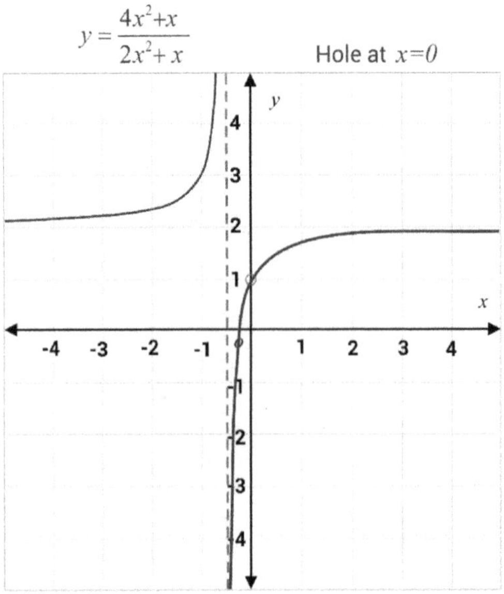

Geometry and Measurement

Plane Geometry

Locations on the plane that have no width or breadth are called *points*. These points usually will be denoted with capital letters such as *P*.

Any pair of points *A*, *B* on the plane will determine a unique straight line between them. This line is denoted *AB*. Sometimes to emphasize a line is being considered, this will be written as $\overleftrightarrow{AB}$.

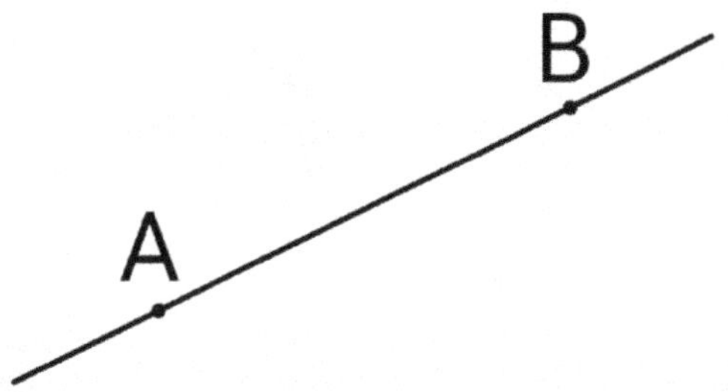

If the Cartesian coordinates for *A* and *B* are known, then the distance $d(A, B)$ along the line between them can be measured using the *Pythagorean formula*, which states that if $A = (x_1, y_1)$ and $B = (x_2, y_2)$, then the distance between them is $d(A, B) = \sqrt{(x_2 - x_1)^2 + (y_2 - y_1)^2}$.

The part of a line that lies between *A* and *B* is called a *line segment*. It has two endpoints, one at *A* and one at *B*. *Rays* also can be formed. Given points *A* and *B*, a *ray* is the portion of a line that starts at one of these points, passes through the other, and keeps on going. Therefore, a ray has a single endpoint, but the other end goes off to infinity.

Given a pair of points *A* and *B*, a circle centered at *A* and passing through *B* can be formed. This is the set of points whose distance from *A* is exactly $d(A, B)$. The radius of this circle will be $d(A, B)$.

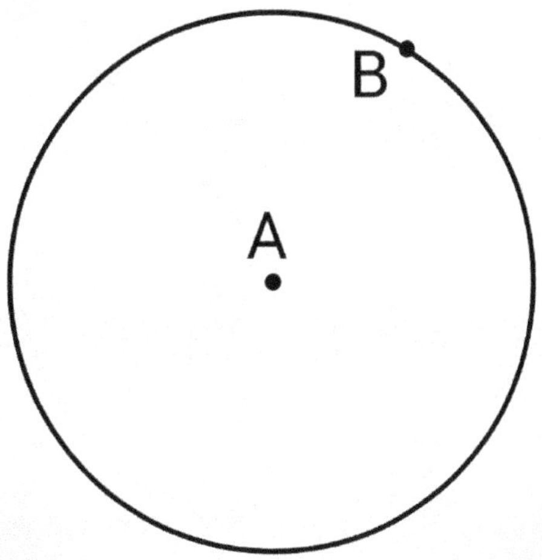

The *circumference* of a circle is the distance traveled by following the edge of the circle for one complete revolution, and the length of the circumference is given by $2\pi r$, where *r* is the radius of the circle. The formula for circumference is $C = 2\pi r$.

When two lines cross, they form an *angle*. The point where the lines cross is called the *vertex* of the angle. The angle can be named by either just using the vertex, $\angle A$, or else by listing three points $\angle BAC$, as shown in the diagram below.

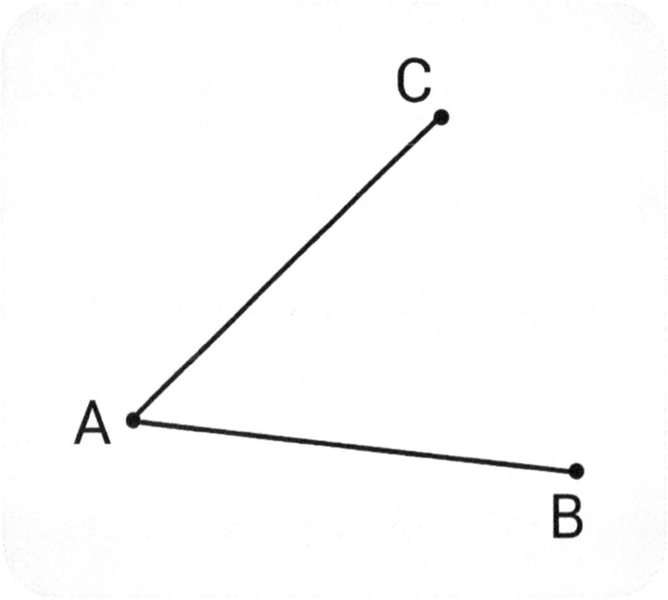

The measurement of an angle can be given in degrees or in radians. In degrees, a full circle is 360 degrees, written 360°. In radians, a full circle is 2π radians.

Given two points on the circumference of a circle, the path along the circle between those points is called an *arc* of the circle. For example, the arc between *B* and *C* is denoted by a thinner line:

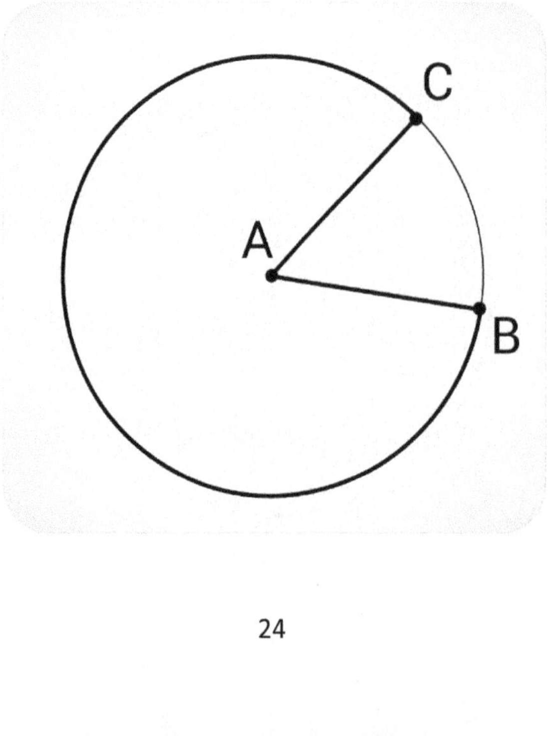

The length of the path along an arc is called the *arc length*. If the circle has radius *r*, then the arc length is given by multiplying the measure of the angle in radians by the radius of the circle.

Two lines are said to be *parallel* if they never intersect. If the lines are *AB* and *CD*, then this is written as $AB \parallel CD$.

If two lines cross to form four quarter-circles, that is, 90° angles, the two lines are *perpendicular*. If the point at which they cross is *B*, and the two lines are *AB* and *BC*, then this is written as $AB \perp BC$.

A *polygon* is a closed figure (meaning it divides the plane into an inside and an outside) consisting of a collection of line segments between points. These points are called the *vertices* of the polygon. These line segments must not overlap one another. Note that the number of sides is equal to the number of angles, or vertices of the polygon. The angles between line segments meeting one another in the polygon are called *interior angles*.

A *regular polygon* is a polygon whose edges are all the same length and whose interior angles are all of equal measure.

A *triangle* is a polygon with three sides. A *quadrilateral* is a polygon with four sides.

A *right triangle* is a triangle that has one 90° angle.

The sum of the interior angles of any triangle must add up to 180°.

An *isosceles triangle* is a triangle in which two of the sides are the same length. In this case, it will always have two congruent interior angles. If a triangle has two congruent interior angles, it will always be isosceles.

An *equilateral triangle* is a triangle whose sides are all the same length and whose angles are all equivalent to one another, equal to 60°. Equilateral triangles are examples of regular polygons. Note that equilateral triangles are also isosceles.

A *rectangle* is a quadrilateral whose interior angles are all 90°. A rectangle has two sets of sides that are equal to one another.

A *square* is a rectangle whose width and height are equal. Therefore, squares are regular polygons.

A *parallelogram* is a quadrilateral in which the opposite sides are parallel and equivalent to each other.

Transformations of a Plane

Given a figure drawn on a plane, many changes can be made to that figure, including *rotation*, *translation*, and *reflection*. Rotations turn the figure about a point, translations slide the figure, and reflections flip the figure over a specified line. When performing these transformations, the original figure is called the *pre-image*, and the figure after transformation is called the *image*.

More specifically, *translation* means that all points in the figure are moved in the same direction by the same distance. In other words, the figure is slid in some fixed direction. Of course, while the entire figure is slid by the same distance, this does not change any of the measurements of the figures involved. The result will have the same distances and angles as the original figure.

In terms of Cartesian coordinates, a translation means a shift of each of the original points (x, y) by a fixed amount in the *x* and *y* directions, to become $(x + a, y + b)$.

Another procedure that can be performed is called *reflection*. To do this, a line in the plane is specified, called the *line of reflection*. Then, take each point and flip it over the line so that it is the same distance from the line but on the opposite side of it. This does not change any of the distances or angles involved, but it does reverse the order in which everything appears.

To reflect something over the *x*-axis, the points (x, y) are sent to $(x, -y)$. To reflect something over the *y*-axis, the points (x, y) are sent to the points $(-x, y)$. Flipping over other lines is not something easy to express in Cartesian coordinates. However, by drawing the figure and the line of reflection, the distance to the line and the original points can be used to find the reflected figure.

Example: Reflect this triangle with vertices (-1, 0), (2, 1), and (2, 0) over the *y*-axis. The pre-image is shown below.

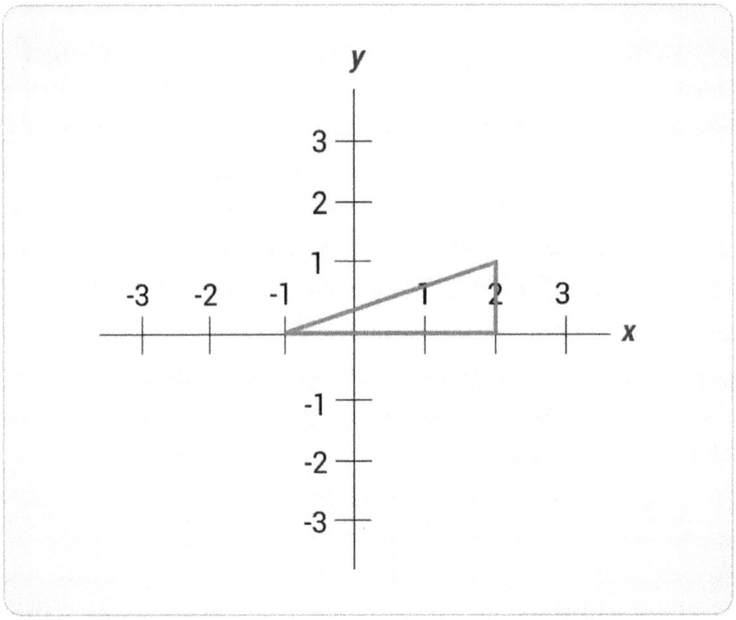

To do this, flip the *x* values of the points involved to the negatives of themselves, while keeping the *y* values the same. The image is shown here.

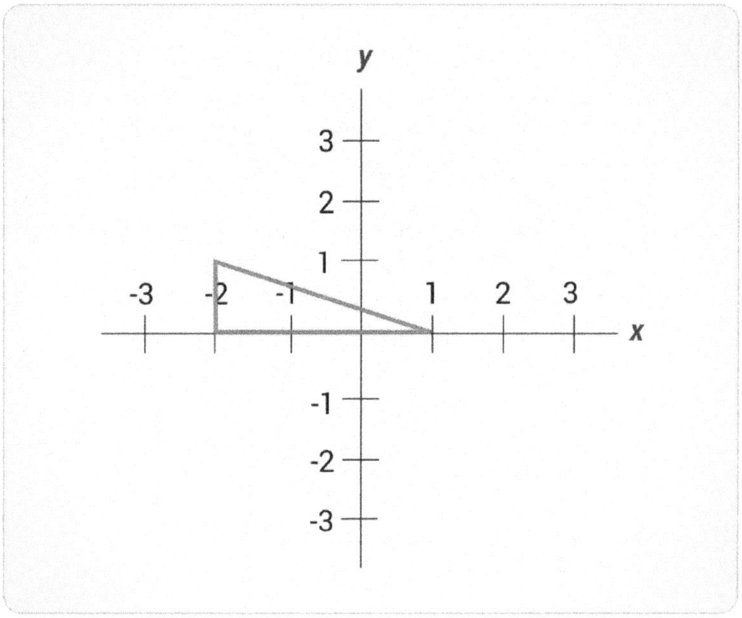

The new vertices will be (1, 0), (-2, 1), and (-2, 0).

Another procedure that does not change the distances and angles in a figure is *rotation*. In this procedure, pick a center point, then rotate every vertex along a circle around that point by the same angle. This procedure is also not easy to express in Cartesian coordinates, and this is not a requirement on this test. However, as with reflections, it's helpful to draw the figures and see what the result of the rotation would look like. This transformation can be performed using a compass and protractor.

Each one of these transformations can be performed on the coordinate plane without changes to the original dimensions or angles.

If two figures in the plane involve the same distances and angles, they are called *congruent figures*. In other words, two figures are congruent when they go from one form to another through reflection, rotation, and translation, or a combination of these.

Remember that rotation and translation will give back a new figure that is identical to the original figure, but reflection will give back a mirror image of it.

To recognize that a figure has undergone a rotation, check to see that the figure has not been changed into a mirror image, but that its orientation has changed (that is, whether the parts of the figure now form different angles with the *x* and *y* axes).

To recognize that a figure has undergone a translation, check to see that the figure has not been changed into a mirror image, and that the orientation remains the same.

To recognize that a figure has undergone a reflection, check to see that the new figure is a mirror image of the old figure.

Keep in mind that sometimes a combination of translations, reflections, and rotations may be performed on a figure.

Dilation

A *dilation* is a transformation that preserves angles, but not distances. This can be thought of as stretching or shrinking a figure. If a dilation makes figures larger, it is called an *enlargement*. If a dilation makes figures smaller, it is called a *reduction*. The easiest example is to dilate around the origin. In this case, multiply the *x* and *y* coordinates by a *scale factor*, k, sending points (x, y) to (kx, ky).

As an example, draw a dilation of the following triangle, whose vertices will be the points (-1, 0), (1, 0), and (1, 1).

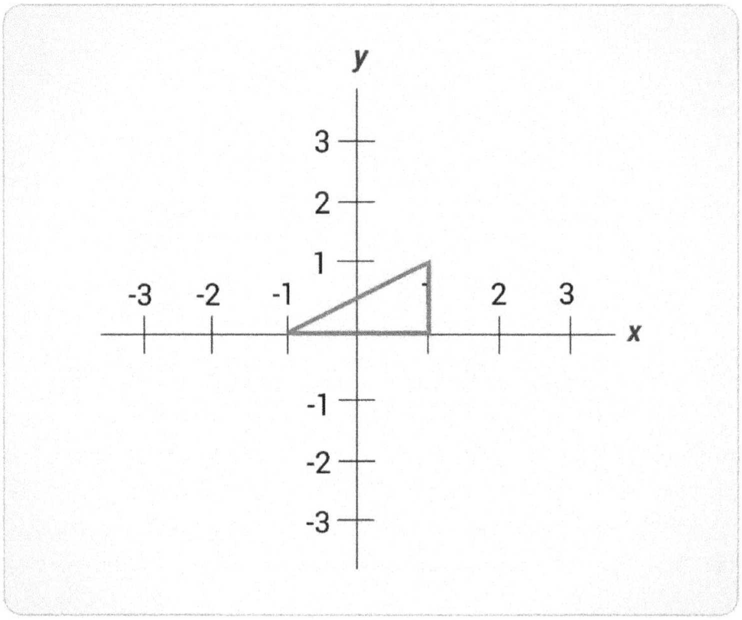

For this problem, dilate by a scale factor of 2, so the new vertices will be (-2, 0), (2, 0), and (2, 2).

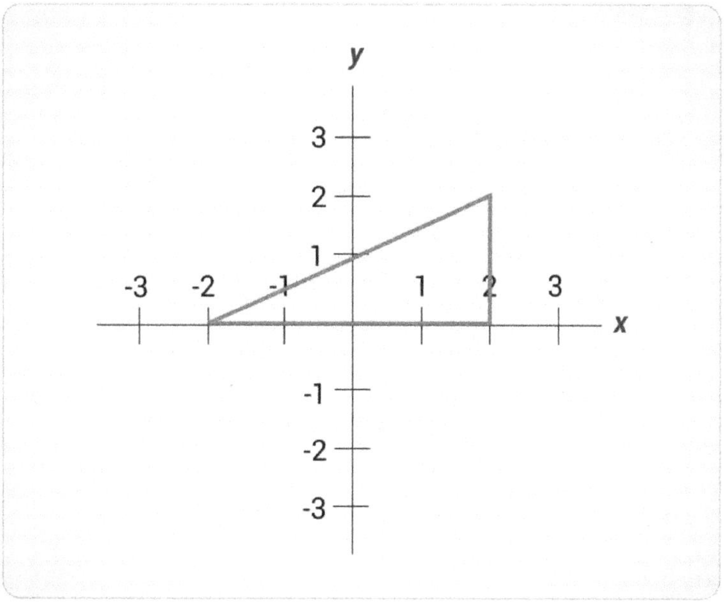

Note that after a dilation, the distances between the vertices of the figure will have changed, but the angles remain the same. The two figures that are obtained by dilation, along with possibly translation, rotation, and reflection, are all *similar* to one another. Another way to think of this is that similar figures have the same number of vertices and edges, and their angles are all the same. Similar figures have the same basic shape, but are different in size.

Symmetry

Using the types of transformations above, if an object can undergo these changes and not appear to have changed, then the figure is symmetrical. If an object can be split in half by a line and flipped over that line to lie directly on top of itself, it is said to have *line symmetry*. An example of both types of figures is seen below.

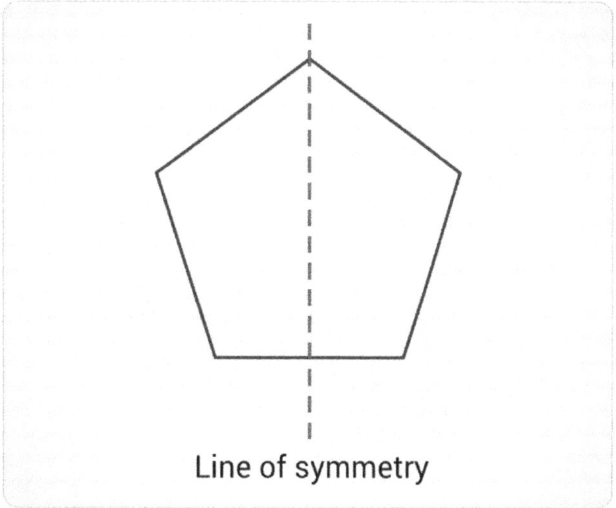

If an object can be rotated about its center to any degree smaller than 360, and it lies directly on top of itself, the object is said to have *rotational symmetry*. An example of this type of symmetry is shown below. The pentagon has an order of 5.

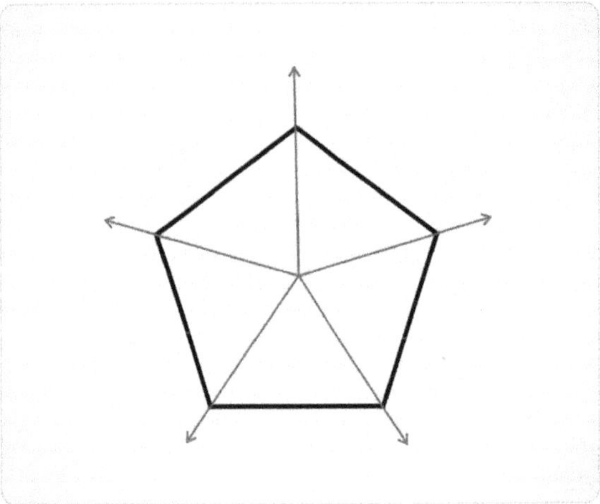

The rotational symmetry lines in the figure above can be used to find the angles formed at the center of the pentagon. Knowing that all of the angles together form a full circle, at 360 degrees, the figure can be split into 5 angles equally. By dividing the 360° by 5, each angle is 72°.

Given the length of one side of the figure, the perimeter of the pentagon can also be found using rotational symmetry. If one side length was 3 cm, that side length can be rotated onto each other side length four times. This would give a total of 5 side lengths equal to 3 cm. To find the perimeter, or distance around the figure, multiply 3 by 5. The perimeter of the figure would be 15 cm.

If a line cannot be drawn anywhere on the object to flip the figure onto itself or rotated less than or equal to 180 degrees to lay on top of itself, the object is asymmetrical. Examples of these types of figures are shown below.

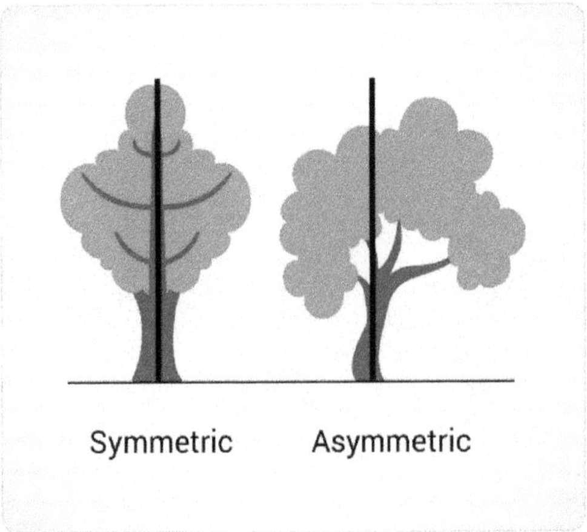

Perimeters and Areas

The *perimeter* of a polygon is the total length of a trip around the whole polygon, starting and ending at the same point. It is found by adding up the lengths of each line segment in the polygon. For a rectangle with sides of length *x* and *y*, the perimeter will be $2x + 2y$.

The area of a polygon is the area of the region that it encloses. Regarding the area of a rectangle with sides of length *x* and *y*, the area is given by xy. For a triangle with a base of length *b* and a height of length *h*, the area is $\frac{1}{2}bh$.

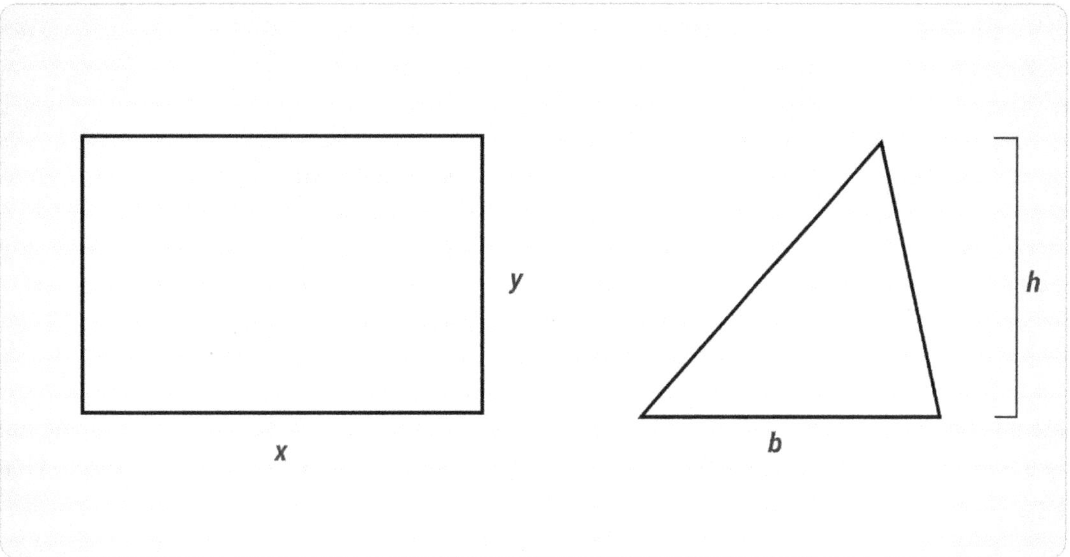

To find the areas of more general polygons, it is usually easiest to break up the polygon into rectangles and triangles. For example, find the area of the following figure whose vertices are (-1, 0), (-1, 2), (1, 3), and (1, 0).

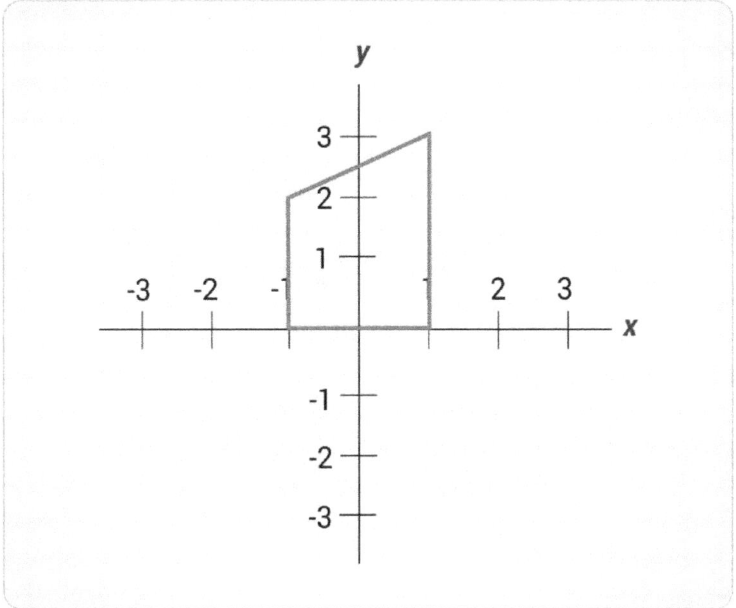

Separate this into a rectangle and a triangle as shown:

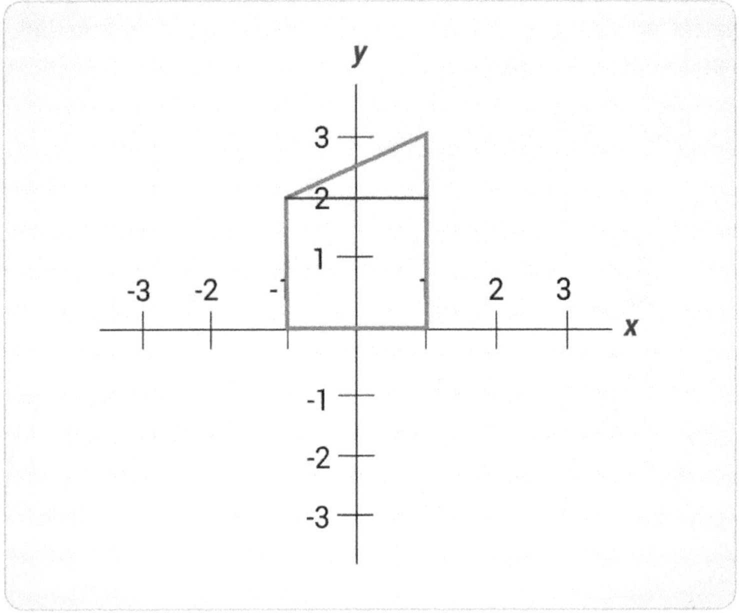

The rectangle has a height of 2 and a width of 2, so it has a total area of $2 \times 2 = 4$. The triangle has a width of 2 and a height of 1, so it has an area of $\frac{1}{2} 2 \times 1 = 1$. Therefore, the entire quadrilateral has an area of 4 + 1 = 5.

As another example, suppose someone wants to tile a rectangular room that is 10 feet by 6 feet using triangular tiles that are 12 inches by 6 inches. How many tiles would be needed? To figure this, first find the area of the room, which will be $10 \times 6 = 60$ square feet. The dimensions of the triangle are 1 foot by ½ foot, so the area of each triangle is $\frac{1}{2} \times 1 \times \frac{1}{2} = \frac{1}{4}$ square feet. Notice that the dimensions of the triangle had to be converted to the same units as the rectangle. Now, take the total area divided by the area of one tile to find the answer: $\frac{60}{\frac{1}{4}} = 60 \times 4 = 240$ tiles required.

Volumes and Surface Areas

Geometry in three dimensions is similar to geometry in two dimensions. The main new feature is that three points now define a unique *plane* that passes through each of them. Three dimensional objects can be made by putting together two dimensional figures in different surfaces. Below, some of the possible three dimensional figures will be provided, along with formulas for their volumes and surface areas.

A rectangular prism is a box whose sides are all rectangles meeting at 90° angles. Such a box has three dimensions: length, width, and height. If the length is *x*, the width is *y*, and the height is *z*, then the volume is given by $V = xyz$.

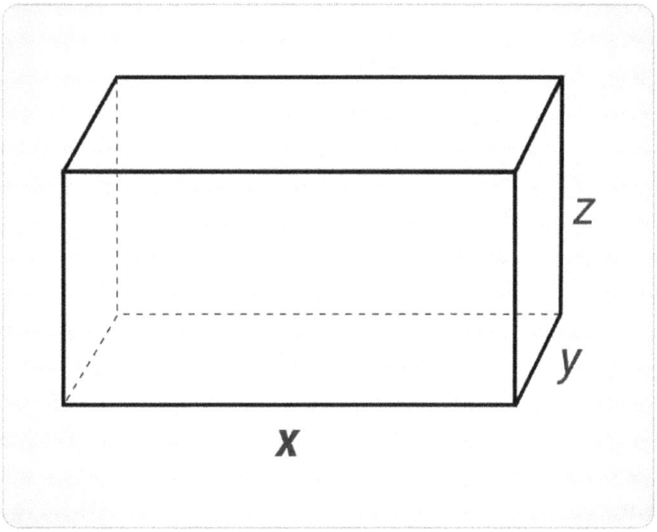

The surface area will be given by computing the surface area of each rectangle and adding them together. There are a total of six rectangles. Two of them have sides of length *x* and *y*, two have sides of length *y* and *z*, and two have sides of length *x* and *z*. Therefore, the total surface area will be given by $SA = 2xy + 2yz + 2xz$.

A *rectangular pyramid* is a figure with a rectangular base and four triangular sides that meet at a single vertex. If the rectangle has sides of length *x* and *y*, then the volume will be given by $V = \frac{1}{3}xyh$.

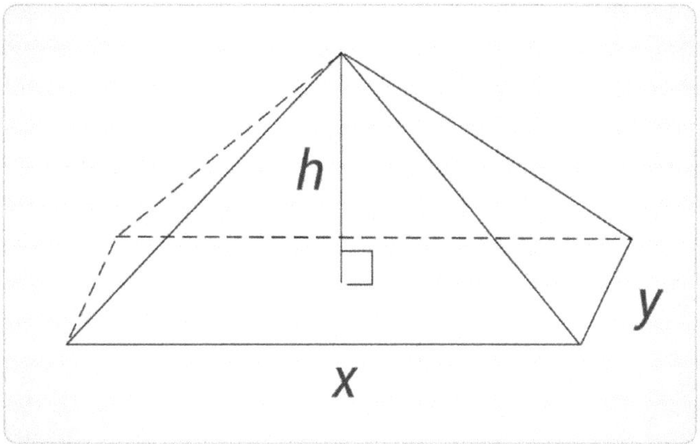

To find the surface area, the dimensions of each triangle need to be known. However, these dimensions can differ depending on the problem in question. Therefore, there is no general formula for calculating total surface area.

A *sphere* is a set of points all of which are equidistant from some central point. It is like a circle, but in three dimensions. The volume of a sphere of radius *r* is given by $V = \frac{4}{3}\pi r^3$. The surface area is given by $A = 4\pi r^2$.

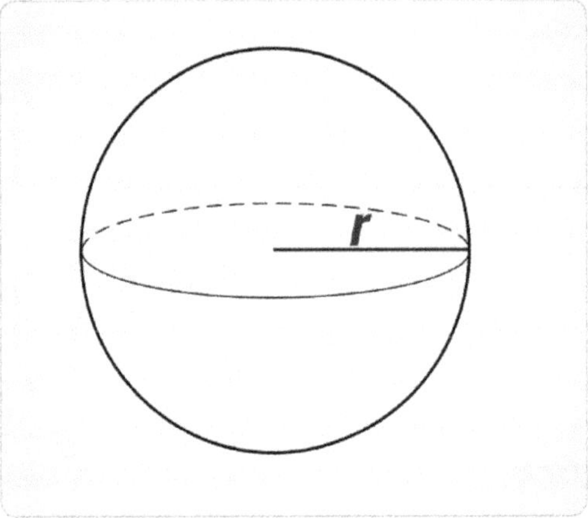

The Pythagorean Theorem

The Pythagorean theorem is an important result in geometry. It states that for right triangles, the sum of the squares of the two shorter sides will be equal to the square of the longest side (also called the *hypotenuse*). The longest side will always be the side opposite to the 90° angle. If this side is called *c*, and the other two sides are *a* and *b*, then the Pythagorean theorem states that $c^2 = a^2 + b^2$. Since lengths are always positive, this also can be written as $c = \sqrt{a^2 + b^2}$.

A diagram to show the parts of a triangle using the Pythagorean theorem is below.

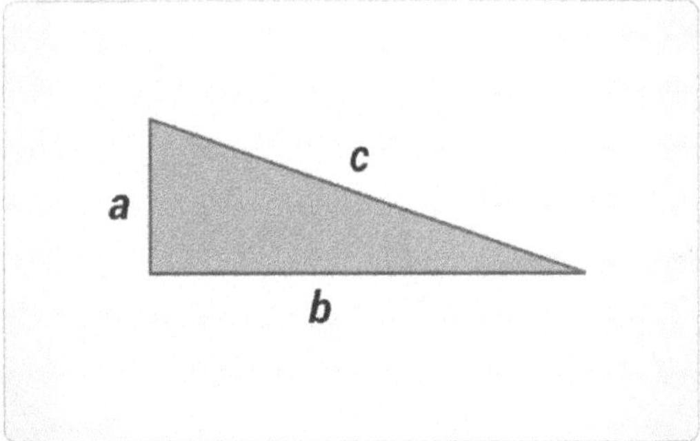

As an example of the theorem, suppose that Shirley has a rectangular field that is 5 feet wide and 12 feet long, and she wants to split it in half using a fence that goes from one corner to the opposite corner. How long will this fence need to be? To figure this out, note that this makes the field into two right triangles, whose hypotenuse will be the fence dividing it in half. Therefore, the fence length will be given by $\sqrt{5^2 + 12^2} = \sqrt{169} = 13$ feet long.

Similar Figures and Proportions

Sometimes, two figures are similar, meaning they have the same basic shape and the same interior angles, but they have different dimensions. If the ratio of two corresponding sides is known, then that ratio, or scale factor, holds true for all of the dimensions of the new figure.

Here is an example of applying this principle. Suppose that Lara is 5 feet tall and is standing 30 feet from the base of a light pole, and her shadow is 6 feet long. How high is the light on the pole? To figure this, it helps to make a sketch of the situation:

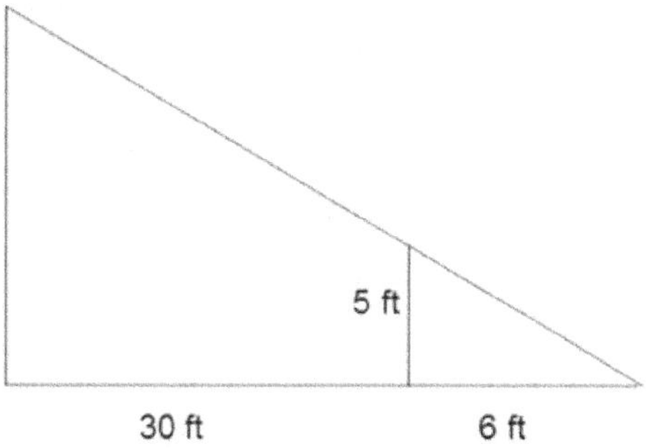

The light pole is the left side of the triangle. Lara is the 5-foot vertical line. Notice that there are two right triangles here, and that they have all the same angles as one another. Therefore, they form similar triangles. So, figure the ratio of proportionality between them.

The bases of these triangles are known. The small triangle, formed by Lara and her shadow, has a base of 6 feet. The large triangle, formed by the light pole along with the line from the base of the pole out to the end of Lara's shadow is $30 + 6 = 36$ feet long. So, the ratio of the big triangle to the little triangle will be $\frac{36}{6} = 6$. The height of the little triangle is 5 feet. Therefore, the height of the big triangle will be $6 \times 5 = 30$ feet, meaning that the light is 30 feet up the pole.

Notice that the perimeter of a figure changes by the ratio of proportionality between two similar figures, but the area changes by the *square* of the ratio. This is because if the length of one side is doubled, the area is quadrupled.

As an example, suppose two rectangles are similar, but the edges of the second rectangle are three times longer than the edges of the first rectangle. The area of the first rectangle is 10 square inches. How much more area does the second rectangle have than the first?

To answer this, note that the area of the second rectangle is $3^2 = 9$ times the area of the first rectangle, which is 10 square inches. Therefore, the area of the second rectangle is going to be $9 \times 10 = 90$ square inches. This means it has $90 - 10 = 80$ square inches more area than the first rectangle.

As a second example, suppose X and Y are similar right triangles. The hypotenuse of X is 4 inches. The area of Y is $\frac{1}{4}$ the area of X. What is the hypotenuse of Y?

First, realize the area has changed by a factor of $\frac{1}{4}$. The area changes by a factor that is the *square* of the ratio of changes in lengths, so the ratio of the lengths is the square root of the ratio of areas. That means that the ratio of lengths must be is $\sqrt{\frac{1}{4}} = \frac{1}{2}$, and the hypotenuse of Y must be $\frac{1}{2} \times 4 = 2$ inches.

Volumes between similar solids change like the cube of the change in the lengths of their edges. Likewise, if the ratio of the volumes between similar solids is known, the ratio between their lengths is known by finding the cube root of the ratio of their volumes.

For example, suppose there are two similar rectangular pyramids X and Y. The base of X is 1 inch by 2 inches, and the volume of X is 8 inches. The volume of Y is 64 inches. What are the dimensions of the base of Y?

To answer this, first find the ratio of the volume of Y to the volume of X. This will be given by $\frac{64}{8} = 8$. Now the ratio of lengths is the cube root of the ratio of volumes, or $\sqrt[3]{8} = 2$. So, the dimensions of the base of Y must be 2 inches by 4 inches.

Data Analysis, Statistics, and Probability

What are Statistics?

The field of statistics describes relationships between quantities that are related, but not necessarily in a deterministic manner. For example, a graduating student's salary will often be higher when the student graduates with a higher GPA, but this is not always the case. Likewise, people who smoke tobacco are more likely to develop lung cancer, but, in fact, it is possible for non-smokers to develop the disease as well. *Statistics* describes these kinds of situations, where the likelihood of some outcome depends on the starting data.

Descriptive statistics involves analyzing a collection of data to describe its broad properties such average (or mean), what percent of the data falls within a given range, and other such properties. An example of this would be taking all of the test scores from a given class and calculating the average test score. *Inferential statistics* attempts to use data about a subset of some population to make inferences about the rest of the population. An example of this would be taking a collection of students who received tutoring and comparing their results to a collection of students who did not receive tutoring, then using that comparison to try to predict whether the tutoring program in question is beneficial.

To be sure that inferences have a high probability of being true for the whole population, the subset that is analyzed needs to resemble a miniature version of the population as closely as possible. For this reason, statisticians like to choose random samples from the population to study, rather than picking a specific group of people based on some similarity. For example, studying the incomes of people who live in Portland does not tell anything useful about the incomes of people who live in Tallahassee.

Mean, Median, and Mode

<u>Mean</u>
Suppose that you have a set of data points and some description of the general properties of this data need to be found.

The first property that can be defined for this set of data is the *mean*. This is the same as average. To find the mean, add up all the data points, then divide by the total number of data points. For example, suppose that in a class of 10 students, the scores on a test were 50, 60, 65, 65, 75, 80, 85, 85, 90, 100. Therefore, the average test score will be:

$$\frac{50 + 60 + 65 + 65 + 75 + 80 + 85 + 85 + 90 + 100}{10} = 75.5$$

The mean is a useful number if the distribution of data is normal (more on this later), which roughly means that the frequency of different outcomes has a single peak and is roughly equally distributed on both sides of that peak. However, it is less useful in some cases where the data might be split or where there are some *outliers*. Outliers are data points that are far from the rest of the data. For example, suppose there are 11 executives and 90 employees at a company. The executives make \$1000 per hour, and the employees make \$10 per hour.

Therefore, the average pay rate will be:

$$\frac{\$1000 \times 11 + \$10 \times 90}{100} = \$119\ per\ hour$$

In this case, this average is not very descriptive since it's not close to the actual pay of the executives *or* the employees.

<u>Median</u>
Another useful measurement is the *median*. In a data set, the median is the point in the middle. The middle refers to the point where half the data comes before it and half comes after, when the data is recorded in numerical order. For instance, these are the speeds of the fastball of a pitcher during the last inning that he pitched (in order from least to greatest):

90, 92, 93, 93, 95, 96, 97, 97, 97

There are nine total numbers, so the middle or *median* number in the 5th one, which is 95.

In cases where the number of data points is an even number, then the average of the two middle points is taken. In the previous example of test scores, the two middle points are 75 and 80. Since there is no single point, the average of these two scores needs to be found. The average is:

$$\frac{75 + 80}{2} = 77.5$$

The median is generally a good value to use if there are a few outliers in the data. It prevents those outliers from affecting the "middle" value as much as when using the mean.

Since an outlier is a data point that is far from most of the other data points in a data set, this means an outlier also is any point that is far from the median of the data set. The outliers can have a substantial effect on the mean of a data set, but usually do not change the median or mode, or do not change them by a large quantity. For example, consider the data set (3, 5, 6, 6, 6, 8). This has a median of 6 and a mode of 6, with a mean of $\frac{34}{6} \approx 5.67$. Now, suppose a new data point of 1000 is added so that the data set is now (3, 5, 6, 6, 6, 8, 1000). This does not change the median or mode, which are both still 6. However, the average is now $\frac{1034}{7}$, which is approximately 147.7. In this case, the median and mode will be better descriptions for most of the data points.

The reason for outliers in a given data set is a complicated problem. It is sometimes the result of an error by the experimenter, but often they are perfectly valid data points that must be taken into consideration.

Mode

One additional measure to define for *X* is the *mode*. This is the data point that appears most frequently. If two or more data points all tie for the most frequent appearance, then each of them is considered a mode. In the case of the test scores, where the numbers were 50, 60, 65, 65, 75, 80, 85, 85, 90, 100, there are two modes: 65 and 85.

Quartiles and Percentiles

The *first quartile* of a set of data *X* refers to the largest value from the first ¼ of the data points. In practice, there are sometimes slightly different definitions that can be used, such as the median of the first half of the data points (excluding the median itself if there are an odd number of data points). The term also has a slightly different use: when it is said that a data point lies *in the first quartile*, it means it is less than or equal to the median of the first half of the data points. Conversely, if it lies *at* the first quartile, then it is equal to the first quartile.

When it is said that a data point lies in the *second quartile*, it means it is between the first quartile and the median.

The *third quartile* refers to data that lies between ½ and ¾ of the way through the data set. Again, there are various methods for defining this precisely, but the simplest way is to include all of the data that lie between the median and the median of the top half of the data.

Data that lies in the *fourth quartile* refers to all of the data above the third quartile.

Percentiles may be defined in a similar manner to quartiles. Generally, this is defined in the following manner:

If a data point lies *in the n-th percentile*, this means it lies in the range of the first *n*% of the data.

If a data point lies *at* the *n*-th percentile, then it means that *n*% of the data lies below this data point.

Standard Deviation

Given a data set *X* consisting of data points $(x_1, x_2, x_3, \dots x_n)$, the *variance* of *X* is defined to be:

$$\frac{\sum_{i=1}^{n}(x_i - \bar{X})^2}{n}$$

This means that the variance of *X* is the average of the squares of the differences between each data point and the mean of *X*. In the formula, $\bar{X}$ is the mean of the values in the data set, and x_i represents each individual value in the data set. The sigma notation indicates that the sum should be found with n being the number of values to add together. $i = 1$ means that the values should begin with the first value.

Given a data set X consisting of data points $(x_1, x_2, x_3, \dots x_n)$, the *standard deviation* of X is defined to be

$$s_x = \sqrt{\frac{\sum_{i=1}^{n}(x_i - \bar{X})^2}{n}}$$

In other words, the standard deviation is the square root of the variance.

Both the variance and the standard deviation are measures of how much the data tend to be spread out. When the standard deviation is low, the data points are mostly clustered around the mean. When the standard deviation is high, this generally indicates that the data are quite spread out, or else that there are a few substantial outliers.

As a simple example, compute the standard deviation for the data set (1, 3, 3, 5). First, compute the mean, which will be $\frac{1+3+3+5}{4} = \frac{12}{4} = 3$. Now, find the variance of X with the formula:

$$\sum_{i=1}^{4}(x_i - \bar{X})^2$$

$$(1-3)^2 + (3-3)^2 + (5-3)^2$$

$$-2^2 + 0^2 + 0^2 + 2^2$$

$$8$$

Therefore, the variance is $\frac{8}{4} = 2$. Taking the square root, the standard deviation will be $\sqrt{2}$.

Note that the standard deviation only depends upon the mean, not upon the median or mode(s). Generally, if there are multiple modes that are far apart from one another, the standard deviation will be high. A high standard deviation does not always mean there are multiple modes, however.

Fitting Functions to Data

Sometimes, when data are measured, it is not simply measuring the frequency of a given outcome, but rather measuring a relationship between two different quantities. In these cases, there is usually one variable that is controlled, the *independent variable*, and one that depends on this variable, the *dependent variable*. If there is a relationship between the two variables, then they are said to be *correlated*.

There are two caveats to these terms. First, the independent variable is not necessarily controlled by the experimenters. It is simply the one chosen to organize the data. In other words, the data are divided up based on the value of an independent variable. Second, finding a significant relationship between the dependent variable and the independent variable does not necessarily imply that there is a causal relationship between the two variables. It only means that once the independent variable is known, a fairly accurate prediction of the dependent variable can be made. This is often expressed with the phrase *correlation does not imply causation*. In other words, just because there is a relationship between two variables does not mean that one is the cause of the other. There could be other factors involved that are the real cause.

Consider some examples. An experimenter could do an experiment in which the independent variable is the number of hours that a student studies for a given test, and the dependent variable is the score the student receives when he or she actually takes the test. Such an experiment would attempt to measure whether there is a relationship between the time spent studying and the score a student receives when taking the test. The expectation would be that the larger value of the independent variable would yield a larger value for the dependent variable. Another experimenter might do an experiment with runners, where the independent variable is the length of the runner's leg, and the dependent variable is the time it takes for the runner to run a fixed distance. In this experiment, as the independent variable increases, the dependent variable would be expected to decrease.

As an example of the phenomenon that correlation does not imply causation, consider an experiment where the independent variable is the value of a person's house, and the dependent variable is their income. Although people in more expensive houses are expected to make more money, it is clear that their expensive houses are not the cause of them making more money. This illustrates one example of why it is important for experimenters to be careful when drawing conclusions about causation from their data.

Linear Data Fitting

The simplest type of correlation between two variables is a *linear correlation*. If the independent variable is x and the dependent variable is y, then a linear correlation means $y = mx + b$. If m is positive, then y will increase as x increases. While if m is negative, then y decreases while x increases. The variable b represents the value of y when x is 0.

As one example of such a correlation, consider a manufacturing plant. Suppose x is the number of units produced by the plant, and y is the cost to the company. In this example, b will be the cost of the plant itself. The plant will cost money even if it is never used, just by buying the machinery. For each unit produced, there will be a cost for the labor and the material. Let m represent this cost to produce one unit of the product.

For a more concrete example, suppose a computer factory costs \$100,000. It requires \$100 of parts and \$50 of labor to make one computer. How much will it cost for a company to make 1000 computers? To figure this, let y be the amount of money the company spends, and let x be the number of computers. The cost of the factory is \$100,000, so $b = 100{,}000$. On the other hand, the cost of producing a computer is the parts plus labor, or \$150, so $m = 150$. Therefore, $y = 150x + 100{,}000$. Substitute 1000 for x and get $y = 150 \times 1000 + 100{,}000 = 150{,}000 + 1000 = 250{,}000$. It will cost the company \$250,000 to make 1000 computers.

Probabilities

Given a set of possible outcomes X, a *probability distribution* on X is a function that assigns a probability to each possible outcome. If the outcomes are $(x_1, x_2, x_3, \ldots x_n)$, and the probability distribution is p, then the following rules are applied.

- $0 \leq p(x_i) \leq 1$, for any i.
- $\sum_{i=1}^{n} p(x_i) = 1$.

In other words, the probability of a given outcome must be between zero and 1, while the total probability must be 1.

If $p(x_i)$ is constant, then this is called a *uniform probability distribution*, and $p(x_i) = \frac{1}{n}$. For example, on a six-sided die, the probability of each of the six outcomes will be $\frac{1}{6}$.

If seeking the probability of an outcome occurring in some specific range *A* of possible outcomes, written $P(A)$, add up the probabilities for each outcome in that range. For example, consider a six-sided die, and figure the probability of getting a 3 or lower when it is rolled. The possible rolls are 1, 2, 3, 4, 5, and 6. So, to get a 3 or lower, a roll of 1, 2, or 3 must be completed. The probabilities of each of these is $\frac{1}{6}$, so add these to get $p(1) + p(2) + p(3) = \frac{1}{6} + \frac{1}{6} + \frac{1}{6} = \frac{1}{2}$.

Conditional Probabilities

An outcome occasionally lies within some range of possibilities *B*, and the probability that the outcomes also lie within some set of possibilities *A* needs to be figured. This is called a *conditional probability*. It is written as $P(A|B)$, which is read "the probability of *A* given *B*." The general formula for computing conditional probabilities is:

$$P(A|B) = \frac{P(A \cap B)}{P(B)}$$

However, when dealing with uniform probability distributions, simplify this a bit. Write $|A|$ to indicate the number of outcomes in *A*. Then, for uniform probability distributions, write $P(A|B) = \frac{|A \cap B|}{|B|}$ (recall that $A \cap B$ means "*A* intersect *B*," and consists of all of the outcomes that lie in both *A* and *B*). This means that all possible outcomes do not need to be known. To see why this formula works, suppose that the set of outcomes *X* is $(x_1, x_2, x_3, \dots x_n)$, so that $|X| = n$. Then, for a uniform probability distribution, $P(A) = \frac{|A|}{n}$. However, this means:

$$(A|B) = \frac{P(A \cap B)}{P(B)} = \frac{\frac{|A \cap B|}{n}}{\frac{|B|}{n}} = \frac{|A \cap B|}{|B|}$$

since the *n*'s cancel out.

For example, suppose a die is rolled and it is known that it will land between 1 and 4. However, how many sides the die has is unknown. Figure the probability that the die is rolled higher than 2. To figure this, $P(3)$ or $P(4)$ does not need to be determined, or any of the other probabilities, since it is known that a fair die has a uniform probability distribution. Therefore, apply the formula $\frac{|A \cap B|}{|B|}$. So, in this case *B* is (1, 2, 3, 4) and $A \cap B$ is (3, 4). Therefore $\frac{|A \cap B|}{|B|} = \frac{2}{4} = \frac{1}{2}$.

Conditional probability is an important concept because, in many situations, the likelihood of one outcome can differ radically depending on how something else comes out. The probability of passing a test given that one has studied all of the material is generally much higher than the probability of passing a test given that one has not studied at all. The probability of a person having heart trouble is much lower if that person exercises regularly. The probability that a college student will graduate is higher when his or her SAT scores are higher, and so on. For this reason, there are many people who are interested in conditional probabilities.

Note that in some practical situations, changing the order of the conditional probabilities can make the outcome very different. For example, the probability that a person with heart trouble has exercised regularly is quite different than the probability that a person who exercises regularly will have heart trouble. The probability of a person receiving a military-only award, given that he or she is or was a soldier, is generally not very high, but the probability that a person being or having been a soldier, given that he or she received a military-only award, is 1.

However, in some cases, the outcomes do not influence one another this way. If the probability of *A* is the same regardless of whether *B* is given; that is, if $P(A|B) = P(A)$, then *A* and *B* are considered *independent*. In this case, $P(A|B) = \frac{P(A \cap B)}{P(B)} = P(A)$, so $P(A \cap B) = P(A)P(B)$. In fact, if $P(A \cap B) = P(A)P(B)$, it can be determined that $P(A|B) = P(A)$ and $P(A|B) = P(B)$ by working backward. Therefore, *B* is also independent of *A*.

An example of something being independent can be seen in rolling dice. In this case, consider a red die and a green die. It is expected that when the dice are rolled, the outcome of the green die should not depend in any way on the outcome of the red die. Or, to take another example, if the same die is rolled repeatedly, then the next number rolled should not depend on which numbers have been rolled previously. Similarly, if a coin is flipped, then the next flip's outcome does not depend on the outcomes of previous flips.

This can sometimes be counter-intuitive, since when rolling a die or flipping a coin, there can be a streak of surprising results. If, however, it is known that the die or coin is fair, then these results are just the result of the fact that over long periods of time, it is very likely that some unlikely streaks of outcomes will occur. Therefore, avoid making the mistake of thinking that when considering a series of independent outcomes, a particular outcome is "due to happen" simply because a surprising series of outcomes has already been seen.

There is a second type of common mistake that people tend to make when reasoning about statistical outcomes: the idea that when something of low probability happens, this is surprising. It would be surprising that something with low probability happened after just one attempt. However, with so much happening all at once, it is easy to see at least something happen in a way that seems to have a very low probability. In fact, a lottery is a good example. The odds of winning a lottery are very small, but the odds that somebody wins the lottery each week are actually fairly high. Therefore, no one should be surprised when some low probability things happen.

Addition Rule

The *addition rule* for probabilities states that the probability of *A* or *B* happening is $P(A \cup B) = P(A) + P(B) - P(A \cap B)$. Note that the subtraction of $P(A \cap B)$ must be performed, or else it would result in double counting any outcomes that lie in both *A* and in *B*. For example, suppose that a 20-sided die is being rolled. Fred bets that the outcome will be greater than 10, while Helen bets that it will be greater than 4 but less than 15. What is the probability that at least one of them is correct?

We apply the rule $P(A \cup B) = P(A) + P(B) - P(A \cap B)$, where *A* is that outcome *x* is in the range $x > 10$, and *B* is that outcome *x* is in the range $4 < x < 15$.

$$P(A) = 10 \times \frac{1}{20} = \frac{1}{2}$$

$$P(B) = 10 \times \frac{1}{20} = \frac{1}{2}$$

$P(A \cap B)$ can be computed by noting that $A \cap B$ means the outcome *x* is in the range $10 < x < 15$, so $P(A \cap B) = 4 \times \frac{1}{20} = \frac{1}{5}$. Therefore:

$$P(A \cup B) = P(A) + P(B) - P(A \cap B) = \frac{1}{2} + \frac{1}{2} - \frac{1}{5} = \frac{4}{5}$$

Note that in this particular example, we could also have directly reasoned about the set of possible outcomes $A \cup B$, by noting that this would mean that *x* must be in the range $5 \leq x$. However, this is not always the case, depending on the given information.

Multiplication Rule

The *multiplication rule* for probabilities states the probability of *A* and *B* both happening is $P(A \cap B) = P(A)P(B|A)$. As an example, suppose that when Jamie wears black pants, there is a ½ probability that she wears a black shirt as well, and that she wears black pants ¾ of the time. What is the probability that she is wearing both a black shirt and black pants?

To figure this, use the above formula, where *A* will be "Jamie is wearing black pants," while *B* will be "Jamie is wearing a black shirt." It is known that $P(A)$ is ¾. It is also known that $P(B|A) = \frac{1}{2}$. Multiplying the two, the probability that she is wearing both black pants and a black shirt is:

$$P(A)P(B|A) = \frac{3}{4} \times \frac{1}{2} = \frac{3}{8}$$

TSI Math Practice Test #1

1. Which of the following inequalities is equivalent to $3 - \frac{1}{2}x \geq 2$?
   a. $x \geq 2$
   b. $x \leq 2$
   c. $x \geq 1$
   d. $x \leq 1$

2. The graph of which function has an x-intercept of -2?
   a. $y = 2x - 3$
   b. $y = 4x + 2$
   c. $y = x^2 + 5x + 6$
   d. $y = -\frac{1}{2} \times 2^x$

3. The table below displays the number of three-year-olds at Kids First Daycare who are potty-trained and those who still wear diapers.

	Potty-trained	**Wear diapers**	
Boys	26	22	48
Girls	34	18	52
	60	40	

What is the probability that a three-year-old girl chosen at random from the school is potty-trained?
   a. 52 percent
   b. 34 percent
   c. 65 percent
   d. 57 percent

4. What is the solution to the following system of equations?

$$x^2 - 2x + y = 8$$
$$x - y = -2$$

   a. $(-2, 3)$
   b. There is no solution.
   c. $(-2, 0)$ $(1, 3)$
   d. $(-2, 0)$ $(3, 5)$

5. An equation for the line passing through the origin and the point $(2, 1)$ is
   a. $y = 2x$
   b. $y = \frac{1}{2}x$
   c. $y = x - 2$
   d. $y = x - 2$

6. A rectangle was formed out of pipe cleaner. Its length was $\frac{1}{2}$ feet and its width was $\frac{11}{2}$ inches. What is its area in square inches?

a. $\frac{11}{4}$ inch^2

b. $\frac{11}{2}$ inch^2

c. 22 inch^2

d. 33 inch^2

7. What type of function is modeled by the values in the following table?

X	f(x)
1	2
2	4
3	8
4	16
5	32

a. Linear
b. Exponential
c. Quadratic
d. Cubic

8. Two cards are drawn from a shuffled deck of 52 cards. What's the probability that both cards are Kings if the first card isn't replaced after it's drawn and is a King?

a. $\frac{1}{169}$

b. $\frac{1}{221}$

c. $\frac{1}{13}$

d. $\frac{4}{13}$

9. Write the expression for three times the sum of twice a number and one minus 6.
a. $2x + 1 - 6$
b. $3x + 1 - 6$
c. $3(x + 1) - 6$
d. $3(2x + 1) - 6$

10. Five of six numbers have a sum of 25. The average of all six numbers is 6. What is the sixth number?
a. 8
b. 10
c. 11
d. 12

11. $(2x - 4y)^2 =$
    a. $4x^2 - 16xy + 16y^2$
    b. $4x^2 - 8xy + 16y^2$
    c. $4x^2 - 16xy - 16y^2$
    d. $2x^2 - 8xy + 8y^2$

12. Which of the following shows the correct result of simplifying the following expression:

$$(7n + 3n^3 + 3) + (8n + 5n^3 + 2n^4)$$

    a. $9n^4 + 15n - 2$
    b. $2n^4 + 5n^3 + 15n - 2$
    c. $9n^4 + 8n^3 + 15n$
    d. $2n^4 + 8n^3 + 15n + 3$

13. What is the slope of this line?

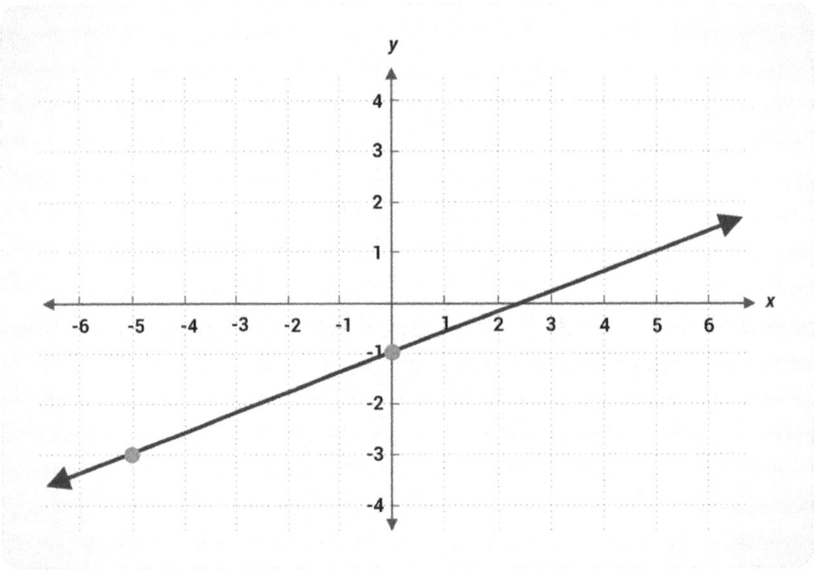

    a. 2
    b. $\frac{5}{2}$
    c. $\frac{1}{2}$
    d. $\frac{2}{5}$

14. What is the product of the following expression?

$$(4x - 8)(5x^2 + x + 6)$$

    a. $20x^3 - 36x^2 + 16x - 48$
    b. $6x^3 - 41x^2 + 12x + 15$
    c. $204 + 11x^2 - 37x - 12$
    d. $2x^3 - 11x^2 - 32x + 20$

15. What is the solution for the following equation?

$$\frac{x^2 + x - 30}{x - 5} = 11$$

a. $x = -6$
b. There is no solution.
c. $x = 16$
d. $x = 5$

16. If x is not zero, then $\frac{3}{x} + \frac{5u}{2x} - \frac{u}{4} =$

a. $\frac{12+10u-ux}{4x}$

b. $\frac{3+5u-ux}{x}$

c. $\frac{12x+10u+ux}{4x}$

d. $\frac{12+10u-u}{4x}$

17. For the following similar triangles, what are the values of x and y (rounded to one decimal place)?

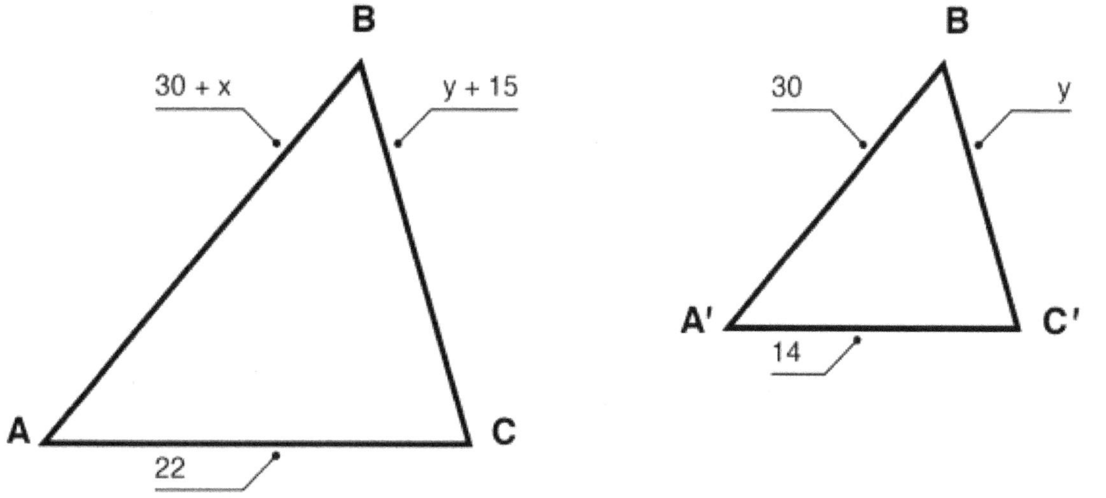

a. $x = 16.5, y = 25.1$
b. $x = 19.5, y = 24.1$
c. $x = 17.1, y = 26.3$
d. $x = 26.3, y = 17.1$

18. What is the perimeter of the figure below? Note that the solid outer line is the perimeter.

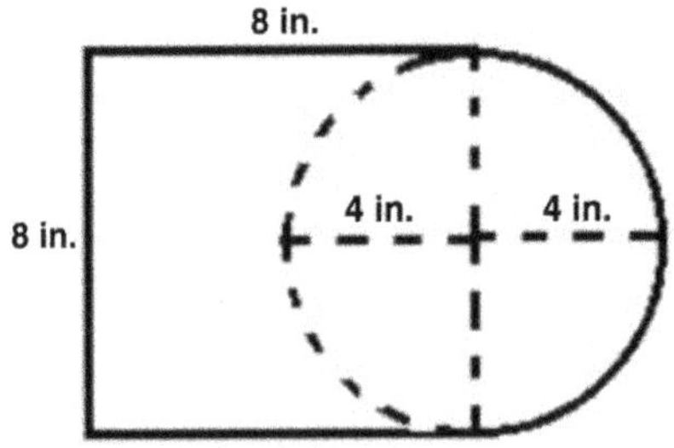

a. 48.565 in
b. 36.565 in
c. 39.78 in
d. 39.565 in

19. A pair of dice is thrown, and the sum of the two scores is calculated. What's the expected value of the roll?

a. 5
b. 6
c. 7
d. 8

20. Karen gets paid a weekly salary and a commission for every sale that she makes. The table below shows the number of sales and her pay for different weeks.

Sales	2	7	4	8
Pay	\$380	\$580	\$460	\$620

Which of the following equations represents Karen's weekly pay?

a. $y = 90x + 200$
b. $y = 90x - 200$
c. $y = 40x + 300$
d. $y = 40x - 300$

Answer Explanations #1

1. B: To simplify this inequality, subtract 3 from both sides to get $-\frac{1}{2}x \geq -1$. Then, multiply both sides by -2 (remembering this flips the direction of the inequality) to get $x \leq 2$.

2. C: An *x*-intercept is the point where the graph crosses the *x*-axis. At this point, the value of *y* is 0. To determine if an equation has an *x*-intercept of -2, substitute -2 for *x,* and calculate the value of *y*. If the value of -2 for *x* corresponds with a *y*-value of 0, then the equation has an *x*-intercept of -2. The only answer choice that produces this result is Choice *C* → $0 = (-2)2 + 5(-2) + 6$.

3. C: The conditional frequency of a girl being potty-trained is calculated by dividing the number of potty-trained girls by the total number of girls: $34 \div 52 = 0.65$. To determine the conditional probability, multiply the conditional frequency by 100: $0.65 \times 100 = 65\%$.

4. D: This system of equations involves one quadratic function and one linear function, as seen from the degree of each equation. One way to solve this is through substitution. Solving for y in the second equation yields $y = x + 2$. Plugging this equation in for the y of the quadratic equation yields $x^2 - 2x + x + 2 = 8$. Simplifying the equation, it becomes $x^2 - x + 2 = 8$. Setting this equal to zero and factoring, it becomes $x^2 - x - 6 = 0 = (x - 3)(x + 2)$. Solving these two factors for x gives the zeros $x = 3, -2$. To find the y-value for the point, each number can be plugged in to either original equation. Solving each one for y yields the points $(3, 5)$ and $(-2, 0)$.

5. B: The slope will be given by $\frac{1-0}{2-0} = \frac{1}{2}$. The *y*-intercept will be 0, since it passes through the origin. Using slope-intercept form, the equation for this line is $y = \frac{1}{2}x$.

6. D: Area = length x width. The answer must be in square inches, so all values must be converted to inches. $\frac{1}{2}$ ft is equal to 6 inches. Therefore, the area of the rectangle is equal to $6 \times \frac{11}{2} = \frac{66}{2} = 33$ square inches.

7. B: The table shows values that are increasing exponentially. The differences between the inputs are the same, while the differences in the outputs are changing by a factor of 2. The values in the table can be modeled by the equation $f(x) = 2^x$.

8. B: For the first card drawn, the probability of a King being pulled is $\frac{4}{52}$. Since this card isn't replaced, if a King is drawn first, the probability of a King being drawn second is $\frac{3}{51}$. The probability of a King being drawn in both the first and second draw is the product of the two probabilities: $\frac{4}{52} \times \frac{3}{51} = \frac{12}{2{,}652}$ which, divided by 12, equals $\frac{1}{221}$.

9. D: The expression is three times the sum of twice a number and 1, which is $3(2x + 1)$. Then, 6 is subtracted from this expression.

10. C: The average is calculated by adding all six numbers, then dividing by 6. The first five numbers have a sum of 25. If the total divided by 6 is equal to 6, then the total itself must be 36. The sixth number must be 36 – 25 = 11.

11. A: To expand a squared binomial, it's necessary use the *First, Inner, Outer, Last Method.* $(2x - 4y)^2 = 2x \times 2x + 2x(-4y) + (-4y)(2x) + (-4y)(-4y) = 4x^2 - 8xy - 8xy + 16y^2 = 4x^2 - 16xy + 16y^2$.

12. D: The expression is simplified by collecting like terms. Terms with the same variable and exponent are like terms, and their coefficients can be added.

13. D: The slope is given by the change in *y* divided by the change in *x*. Specifically, it's:

$$slope = \frac{y_2 - y_1}{x_2 - x_1}$$

The first point is (-5,-3) and the second point is (0,-1). Work from left to right when identifying coordinates. Thus, the point on the left is point 1 (-5,-3) and the point on the right is point 2 (0,-1).

Now we need to just plug those numbers into the equation:

$$slope = \frac{-1 - (-3)}{0 - (-5)}$$

It can be simplified to:

$$slope = \frac{-1 + 3}{0 + 5}$$

$$slope = \frac{2}{5}$$

14. A: Finding the product means distributing one polynomial to the other so that each term in the first is multiplied by each term in the second. Then, like terms can be collected. Multiplying the factors yields the expression $20x^3 + 4x^2 + 24x - 40x^2 - 8x - 48$. Collecting like terms means adding the x^2 terms and adding the x terms. The final answer after simplifying the expression is $20x^3 - 36x^2 + 16x - 48$.

15. B: The equation can be solved by factoring the numerator into $(x + 6)(x - 5)$. Since that same factor $(x - 5)$ exists on top and bottom, that factor cancels. This leaves the equation $x + 6 = 11$. Solving the equation gives the answer $x = 5$. When this value is plugged into the equation, it yields a zero in the denominator of the fraction. Since this is undefined, there is no solution.

16. C: The common denominator here will be 4*x*. Rewrite these fractions as:

$$\frac{3}{x} + \frac{5u}{2x} - \frac{u}{4} = \frac{12}{4x} + \frac{10u}{4x} - \frac{ux}{4x} = \frac{12x + 10u + ux}{4x}$$

17. C: Because the triangles are similar, the lengths of the corresponding sides are proportional. Therefore:

$$\frac{30 + x}{30} = \frac{22}{14} = \frac{y + 15}{y}$$

This results in the equation $14(30 + x) = 22 \cdot 30$ which, when solved, gives $x = 17.1$. The proportion also results in the equation $14(y + 15) = 22y$ which, when solved, gives $y = 26.3$.

18. B: The figure is composed of three sides of a square and a semicircle. The sides of the square are simply added: 8 + 8 + 8 = 24 inches. The circumference of a circle is found by the equation C = 2πr. The radius is 4 in, so the circumference of the circle is 25.13 in. Only half of the circle makes up the outer border of the figure (part of the perimeter) so half of 25.13 in is 12.565 in. Therefore, the total perimeter is: 24 in + 12.565 in = 36.565 in. The other answer choices use the incorrect formula or fail to include all of the necessary sides.

19. C: The expected value is equal to the total sum of each product of individual score and probability. There are 36 possible rolls. The probability of rolling a 2 is $\frac{1}{36}$. The probability of rolling a 3 is $\frac{2}{36}$. The probability of rolling a 4 is $\frac{3}{36}$. The probability of rolling a 5 is $\frac{4}{36}$. The probability of rolling a 6 is $\frac{5}{36}$. The probability of rolling a 7 is $\frac{6}{36}$. The probability of rolling an 8 is $\frac{5}{36}$. The probability of rolling a 9 is $\frac{4}{36}$. The probability of rolling a 10 is $\frac{3}{36}$. The probability of rolling an 11 is $\frac{2}{36}$. Finally, the probability of rolling a 12 is $\frac{1}{36}$.

Each possible outcome is multiplied by the probability of it occurring. Like this:

$$2 \times \frac{1}{36} = a$$

$$3 \times \frac{2}{36} = b$$

$$4 \times \frac{3}{36} = c$$

And so forth.

Then all of those results are added together:

$$a + b + c \ldots = expected\ value$$

In this case, it equals 7.

20. C: $y = 40x + 300$. In this scenario, the variables are the number of sales and Karen's weekly pay. The weekly pay depends on the number of sales. Therefore, weekly pay is the dependent variable (*y*) and the number of sales is the independent variable (*x*). Each pair of values from the table can be written as an ordered pair (*x*, *y*): (2,380), (7,580), (4,460), (8,620). The ordered pairs can be substituted into the equations to see which creates true statements (both sides equal) for each pair. Even if one ordered pair produces equal values for a given equation, the other three ordered pairs must be checked. The only equation which is true for all four ordered pairs is $y = 40x + 300$:

$$380 = 40(2) + 300 \rightarrow 380 = 380$$

$$580 = 40(7) + 300 \rightarrow 580 = 580$$

$$460 = 40(4) + 300 \rightarrow 460 = 460$$

$$620 = 40(8) + 300 \rightarrow 620 = 620$$

TSI Math Practice Test #2

1. The square and circle have the same center. The circle has a radius of *r*. What is the area of the shaded region?

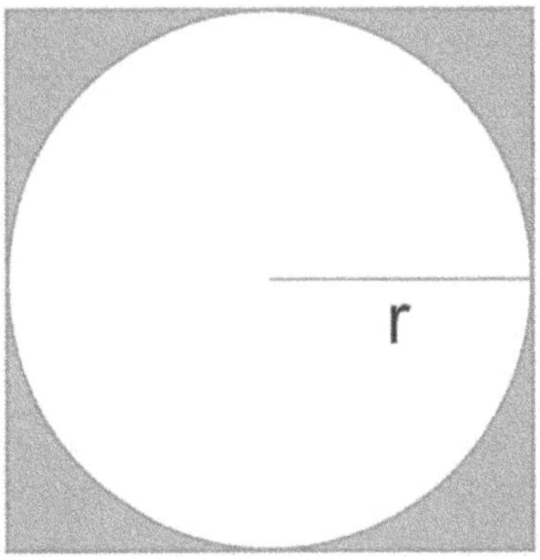

a. $r^2 - \pi r^2$
b. $4r^2 - 2\pi r$
c. $(4 - \pi)r^2$
d. $(\pi - 1)r^2$

2. Solve for X: $\frac{2x}{5} - 1 = 59.$
a. 60
b. 145
c. 150
d. 115

3. At the store, Jan spends $90 on apples and oranges. Apples cost $1 each and oranges cost $2 each. If Jan buys the same number of apples as oranges, how many oranges did she buy?
a. 20
b. 25
c. 30
d. 35

4. Katie works at a clothing company and sold 192 shirts over the weekend. $^1/_3$ of the shirts that were sold were patterned, and the rest were solid. Which mathematical expression would calculate the number of solid shirts Katie sold over the weekend?
a. $192 \times \frac{1}{3}$

b. $192 \div \frac{1}{3}$

c. $192 \times (1 - \frac{1}{3})$

d. $192 \div 3$

5. Given the value of a given stock at monthly intervals, which graph should be used to best represent the trend of the stock?

a. Box plot
b. Line plot
c. Line graph
d. Circle graph

6.

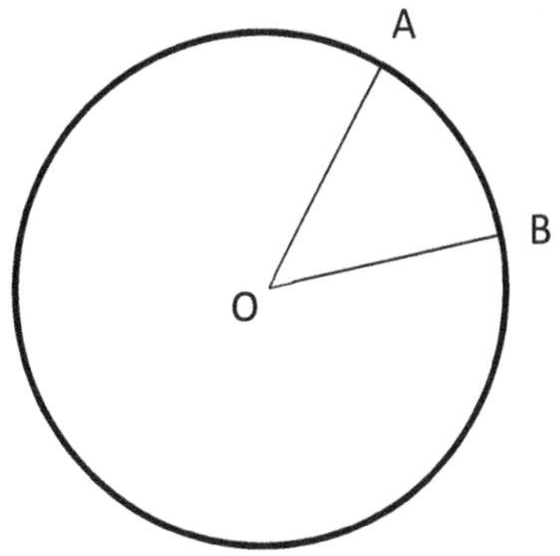

The length of arc $AB = 3\pi$ cm. The length of $\overline{OA} = 12$ cm. What is the degree measure of $\angle AOB$?

a. 30 degrees
b. 40 degrees
c. 45 degrees
d. 55 degrees

7. How will the following algebraic expression be simplified: $(5x^2 - 3x + 4) - (2x^2 - 7)$?

a. x^5
b. $3x^2 - 3x + 11$
c. $3x^2 - 3x - 3$
d. $x - 3$

8. If $-3(x + 4) \geq x + 8$, what is the value of x?

a. $x = 4$
b. $x \geq 2$
c. $x \geq -5$
d. $x \leq -5$

9. For a group of 20 men, the median weight is 180 pounds and the range is 30 pounds. If each man gains 10 pounds, which of the following would be true?

a. The median weight will increase, and the range will remain the same.
b. The median weight and range will both remain the same.
c. The median weight will stay the same, and the range will increase.
d. The median weight and range will both increase.

10. If the ordered pair $(-3, -4)$ is reflected over the x-axis, what's the new ordered pair?

a. $(-3, -4)$
b. $(3, -4)$
c. $(3, 4)$
d. $(-3, 4)$

11. If the volume of a sphere is 288π cubic meters, what are the radius and surface area of the same sphere?

a. Radius 6 meters and surface area 144π square meters
b. Radius 36 meters and surface area 144π square meterc. Radius 6 meters and surface area 12π square meters
d. Radius 36 meters and surface area 12π square meters

12. Ten students take a test. Five students get a 50. Four students get a 70. If the average score is 55, what was the last student's score?

a. 20
b. 40
c. 50
d. 60

13. Which of the following equations best represents the problem below?
The width of a rectangle is 2 centimeters less than the length. If the perimeter of the rectangle is 44 centimeters, then what are the dimensions of the rectangle?

a. $2l + 2(l - 2) = 44$
b. $(l + 2) + (l + 2) + l = 48$
c. $l \times (l - 2) = 44$
d. $(l + 2) + (l + 2) + l = 44$

14. What are the y-intercept(s) for $y = x^2 + 3x - 4$?

a. $y = 1$
b. $y = -4$
c. $y = 3$
d. $y = 4$

15. Which equation is not a function?
  a. $y = |x|$
  b. $y = \sqrt{x}$
  c. $x = 3$
  d. $y = 4$

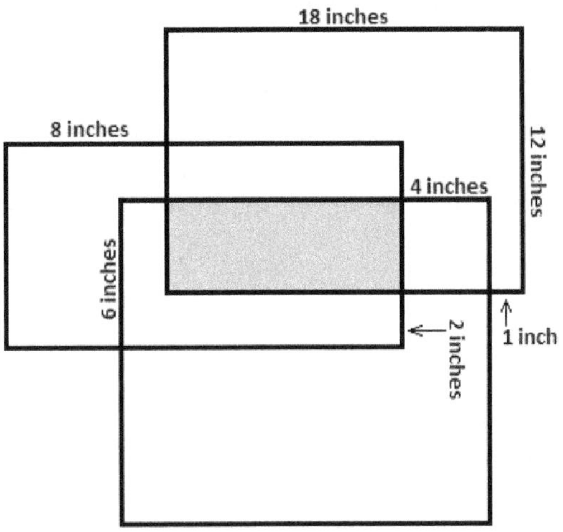

16. In the figure above, what is the area of the shaded region?
  a. 48 sq. inches
  b. 52 sq. inches
  c. 44 sq. inches
  d. 56 sq. inches

17. If $3x = 6y = -2z = 24$, then what does $4xy + z$ equal?
  a. 116
  b. 130
  c. 84
  d. 108

18.

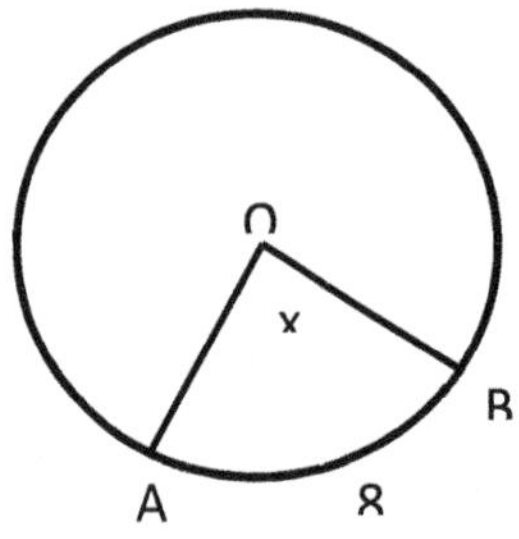

The area of circle O is 49π m. What is the area of the sector formed by $\angle AOB$?

a. 80π m

b. 10.9π m

c. 4.9π m

d. 10π m

19. What is the simplified quotient of the following equation?

$$\frac{5x^3}{3x^2y} \div \frac{25}{3y^9}$$

a. $\frac{125x}{9y^{10}}$

b. $\frac{x}{5y^8}$

c. $\frac{5}{xy^8}$

d. $\frac{xy^8}{5}$

20. Which of the following is a factor of both $x^2 + 4x + 4$ and $x^2 - x - 6$?

a. $x - 3$

b. $x + 2$

c. $x - 2$

d. $x + 3$

Answer Explanations #2

1. C: The area of the shaded region is the area of the square, minus the area of the circle. The area of the circle will be πr^2. The side of the square will be $2r$, so the area of the square will be $4r^2$. Therefore, the difference is $4r^2 - \pi r^2 = (4 - \pi)r^2$.

2. C: X = 150

Set up the initial equation.

$$\frac{2X}{5} - 1 = 59$$

Add 1 to both sides.

$$\frac{2X}{5} - 1 + 1 = 59 + 1$$

Multiply both sides by 5/2.

$$\frac{2X}{5} \times \frac{5}{2} = 60 \times \frac{5}{2} = 150$$

$$X = 150$$

3. C: One apple/orange pair costs $3 total. Therefore, Jan bought 90/3 = 30 total pairs, and hence, she bought 30 oranges.

4. C: $\frac{1}{3}$ of the shirts sold were patterned. Therefore, $1 - \frac{1}{3} = \frac{2}{3}$ of the shirts sold were solid. Anytime "of" a quantity appears in a word problem, multiplication should be used. Therefore, $192 \times \frac{2}{3} = \frac{192 \times 2}{3} = \frac{384}{3} = 128$ solid shirts were sold. The entire expression is $192 \times \left(1 - \frac{1}{3}\right)$.

5. C: Line graph. The scenario involves data consisting of two variables, month, and stock value. Box plots display data consisting of values for one variable. Therefore, a box plot is not an appropriate choice. Both line plots and circle graphs are used to display frequencies within categorical data. Neither can be used for the given scenario. Line graphs display two numerical variables on a coordinate grid and show trends among the variables.

6. C: The formula to find arc length is $s = \theta r$ where *s* is the arc length, θ is the radian measure of the central angle, and *r* is the radius of the circle. Substituting the given information produces $3\pi \text{ cm} = \theta 12 \text{ cm}$. Solving for θ yields $\theta = \frac{\pi}{4}$. To convert from radian to degrees, multiply the radian measure by

$$\frac{180}{\pi}: \frac{\pi}{4} \times \frac{180}{\pi} = 45^{\circ}$$

7. B: $3x^2 - 3x + 11$. By distributing the implied one in front of the first set of parentheses and the – 1 in front of the second set of parentheses, the parenthesis can be eliminated:

$$1(5x^2 - 3x + 4) - 1(2x^2 - 7) = 5x^2 - 3x + 4 - 2x^2 + 7$$

Next, like terms (same variables with same exponents) are combined by adding the coefficients and keeping the variables and their powers the same: $5x^2 - 3x + 4 - 2x^2 + 7 = 3x^2 - 3x + 11$.

8. D: $x \leq -5$. When solving a linear equation or inequality:

Distribution is performed if necessary: $-3(x + 4) \rightarrow -3x - 12 \geq x + 8$. This means that any like terms on the same side of the equation/inequality are combined.

The equation/inequality is manipulated to get the variable on one side. In this case, subtracting x from both sides produces $-4x - 12 \geq 8$.

The variable is isolated using inverse operations to undo addition/subtraction. Adding 12 to both sides produces $-4x \geq 20$.

The variable is isolated using inverse operations to undo multiplication/division. Remember if dividing by a negative number, the relationship of the inequality reverses, so the sign is flipped. In this case, dividing by -4 on both sides produces $x \leq -5$.

9. A: If each man gains 10 pounds, every original data point will increase by 10 pounds. Therefore, the man with the original median will still have the median value, but that value will increase by 10. The smallest value and largest value will also increase by 10 and, therefore, the difference between the two won't change. The range does not change in value and, thus, remains the same.

10. D: When an ordered pair is reflected over an axis, the sign of one of the coordinates must change. When it's reflected over the x-axis, the sign of the y coordinate must change. The x value remains the same. Therefore, the new ordered pair is $(-3, 4)$.

11. A: Because the volume of the given sphere is 288π cubic meters, this gives ${}^4/_3\,\pi r^3 = 288\pi$. This equation is solved for r to obtain a radius of 6 meters. The formula for surface area is $4\pi r^2$ so $SA = 4\pi 6^2 = 144\pi$ square meters.

12. A: Let the unknown score be x. The average will be $\frac{5 \cdot 50 + 4 \cdot 70 + x}{10} = \frac{530 + x}{10} = 55$. Multiply both sides by 10 to get $530 + x = 550$, or $x = 20$.

13. A: The first step is to determine the unknown, which is in terms of the length, l.

The second step is to translate the problem into the equation using the perimeter of a rectangle, $P = 2l + 2w$. The width is the length minus 2 centimeters. The resulting equation is $2l + 2(l - 2) = 44$. The equation can be solved as follows:

$2l + 2l - 4 = 44$	Apply the distributive property on the left side of the equation
$4l - 4 = 44$	Combine like terms on the left side of the equation
$4l = 48$	Add 4 to both sides of the equation
$l = 12$	Divide both sides of the equation by 4

The length of the rectangle is 12 centimeters. The width is the length minus 2 centimeters, which is 10 centimeters. Checking the answers for length and width forms the following equation:

$$44 = 2(12) + 2(10)$$

The equation can be solved using the order of operations to form a true statement: $44 = 44$.

14. B: The y-intercept of an equation is found where the x-value is zero. Plugging zero into the equation for x allows the first two terms to cancel out, leaving -4.

15. C: The equation $x = 3$ is not a function because it does not pass the vertical line test. This test is made from the definition of a function, where each x-value must be mapped to one, and only one, y-value. This equation is a vertical line, so the x-value of 3 is mapped with an infinite number of y-values.

16. B: This can be determined by finding the length and width of the shaded region. The length can be found using the length of the top rectangle which is 18 inches, then subtracting the extra length of 4 inches and 1 inch. This means the length of the shaded region is 13 inches. Next, the width can be determined using the 6 inch measurement and subtracting the 2 inch measurement. This means that the width is 4 inches. Thus, the area is $13 \times 4 = 52$ sq. inches.

17. A: First solve for x, y, and z. So, $3x = 24, x = 8$, $6y = 24, y = 4$, and $-2z = 24, z = -12$. This means the equation would be $4(8)(4) + (-12)$, which equals 116.

18. B: Given the area of the circle, the radius can be found using the formula $A = \pi r^2$. In this case, $49\pi = \pi r^2$, which yields $r = 7$ m. A central angle is equal to the degree measure of the arc it inscribes; therefore, $\angle x = 80°$. The area of a sector can be found using the formula $A = \frac{\theta}{360^0} \times \pi r^2$. In this case, $A = \frac{80^0}{360^0} \times \pi(7)r^2 = 10.9\pi$ m.

19. D: Dividing rational expressions follows the same rule as dividing fractions. The division is changed to multiplication, and the reciprocal is found in the second fraction. This turns the expression into $\frac{5x^3}{3x^2} * \frac{3y^9}{25}$. Multiplying across and simplifying, the final expression is $\frac{xy^8}{5}$.

20. B: To factor $x^2 + 4x + 4$, the numbers needed are those that add to 4 and multiply to 4. Therefore, both numbers must be 2, and the expression factors to $x^2 + 4x + 4 = (x + 2)^2$. Similarly, the expression factors to $x^2 - x - 6 = (x - 3)(x + 2)$, so that they have $x + 2$ in common.

Greetings!

First, we would like to give a huge "thank you" for choosing us and this study guide for your TSI exam. We hope that it will lead you to success on this exam and for your years to come.

Our team has tried to make your preparations as thorough as possible by covering all of the topics you should be expected to know. In addition, our writers attempted to create practice questions identical to what you will see on the day of your actual test. We have also included many test-taking strategies to help you learn the material, maintain the knowledge, and take the test with confidence.

We strive for excellence in our products, and if you have any comments or concerns over the quality of something in this study guide, please send us an email so that we may improve.

As you continue forward in life, we would like to remain alongside you with other books and study guides in our library, such as:

ACCUPLACER: amazon.com/dp/1628455160

We are continually producing and updating study guides in several different subjects. If you are looking for something in particular, all of our products are available on Amazon. You may also send us an email!

Sincerely,
APEX Test Prep
info@apexprep.com

FREE Test Taking Tips DVD Offer

To help us better serve you, we have developed a Test Taking Tips DVD that we would like to give you for FREE. **This DVD covers world-class test taking tips that you can use to be even more successful when you are taking your test.**

All that we ask is that you email us your feedback about your study guide. Please let us know what you thought about it – whether that is good, bad or indifferent.

To get your **FREE Test Taking Tips DVD**, email freedvd@studyguideteam.com with "FREE DVD" in the subject line and the following information in the body of the email:

a. The title of your study guide.

b. Your product rating on a scale of 1-5, with 5 being the highest rating.

c. Your feedback about the study guide. What did you think of it?

d. Your full name and shipping address to send your free DVD.

If you have any questions or concerns, please don't hesitate to contact us at freedvd@studyguideteam.com.

Thanks again!

CPSIA information can be obtained
at www.ICGtesting.com
Printed in the USA
BVHW060146220421
605394BV00008B/1436

9 781628 455953